Risk and Reliability

Risk and Reliability
Coastal and Hydraulic Engineering

Dominic Reeve

Taylor & Francis Group

LONDON AND NEW YORK

First published 2010
by Spon Press

2 Park Square, Milton Park, Abingdon, Oxfordshire OX14 4RN
52 Vanderbilt Avenue, New York, NY 10017

Routledge is an imprint of the Taylor & Francis Group, an informa business

First issued in paperback 2019

Typeset in Sabon by
Integra Software Services Pvt. Ltd.

British Library Cataloguing in Publication Data
A catalogue record for this book is available
from the British Library

Library of Congress Cataloging in Publication Data
Reeve, Dominic.
   Risk and reliability : coastal and hydraulic engineering /
Dominic Reeve.
      p. cm.
   Includes bibliographical references and index.
   1. Coastal engineering.   2. Hydraulic engineering.   I. Title.
   TC205.R445 2009
   627—dc22   2009007962

ISBN13: 978-0-415-46755-1 (hbk)
ISBN13: 978-0-367-38522-4 (pbk)

Dedicated to

Audrey, Sophie and Sasha

Contents

Foreword

Designing infrastructure in the presence of uncertainty has been a central feature of engineering design for many years. The fundamental concepts of probability and risk have been widely applied, but also well disguised in what historically has been an essentially deterministic approach to design. However, in recent years the explicit need to fully recognise risk and its consequences on the design process has brought about a revolution. An essentially deterministic approach has made way for the inevitably probabilistic designs that are now current. This book deliberately sets out to target the application of the important concepts in risk and reliability theory to the problems of coastal and hydraulic engineering. It is a seminal attempt to redress the balance in engineering practice, where the tradition of such concepts being applied in the structural engineering field has been extended into the coastal domain. The book is targeted at both undergraduate and postgraduate audiences, and will likely find a significant readership in the professions. This authoritative text represents a substantial step-change in the treatment of probabilistic processes in the coastal domain at a time of such significant climatic uncertainty.

Professor I. D. Cluckie, FREng, FRSA
Pro-Vice Chancellor (Science and Engineering),
Swansea University
Chairman of the EPSRC Flood Risk Management
Research Consortium (FRMRC)

Swansea
March, 2009

Preface

The concept of balancing risk and cost is central to engineering. It is rarely up to the engineer alone to decide where and how this balance is set. However, engineers have an important role in quantifying, as clearly as possible, the facts that inform financial and political decision-making. The subject of flooding and the design of flood defences is just one such area. Indeed, at the time this book was written some notable flood events had been in the national and international headlines:

- The Asian tsunami caused by an undersea earthquake that occurred at 00:58:53 UTC on 26 December 2004, with an epicentre off the west coast of Sumatra, Indonesia, wreaked death and damage in Indonesia, Thailand and Sri Lanka. Its effects were felt as far away as the east coast of Africa and were even detected in tide gauge records around the UK.
- The village of Boscastle in Devon was the scene of flash flooding resulting from a very intense and concentrated cloud burst on 16 August 2004. Damage was made worse due to the damming effect of a low bridge across the river leading to the sea. As a result of the flooding, the bridge was reconstructed to allow a greater flow of water. This was probably what prevented worse flooding than occurred in Boscastle again on 21 June 2007, when another unusually intense rainfall event affected the village.
- On 29 August 2005, New Orleans sustained major flooding as Hurricane Katrina made landfall very close by, causing severe damage and disruption to the whole of the city. While levees have been rebuilt, and some of the city has returned towards normal, at the time of writing in 2008, large segments of the city remain unrepaired.
- On 25 June 2007, the city of Hull and much of South Yorkshire received over 2 months of average rainfall in 24 hours, leading to prolonged flooding, damage, disruption and displacement of the populace. A similar event affected the county of Gloucester on 20 July 2007, leading to similar results. It has been estimated that over 17,000 people were relocated from their homes as a result of the floods in

summer 2007, and it took more than two years for many to be able to return to their homes from temporary accommodation.

The problems arising from flooding are not new, but recent events such as those above have stimulated public interest in flood defence. In turn, this has raised the profile of flood engineering within the curricula of undergraduate and postgraduate degree programmes. This book is based, in part, on modules in coastal and hydraulic engineering developed over several years in the School of Engineering at the University of Plymouth. It is also influenced by the author's experience of using probabilistic design methods and reliability theory in a range of practical engineering design problems.

Probabilistic or risk-based methods allow uncertainties to be quantified. A probabilistic approach to the design of flood and sea defences has been promoted in the UK by the Department of Environment, Food and Regional Affairs through Project Appraisal Guidance Notes, which define the procedures that applications for grant aid towards flood defence and erosion protection works must follow. The purpose of such an approach is twofold. First, it requires planners and designers to consider carefully the uncertainties inherent in their scheme, and to quantify these as best they can. Second, where good information about the construction materials and the loadings that are to be experienced by the structure are available, a probabilistic approach may assist in reducing uncertainty, thereby lessening conservatism in design and thus improving cost efficiency. As a result, probabilistic methods appear increasingly in the guidance for coast protection and flood defence design.

One difficulty for those wishing to learn and use the techniques of reliability analysis is that it involves topics that are often not standard in current undergraduate engineering courses. This means that the subject can appear difficult and off-putting to the novice. Inevitably, some level of prior knowledge has to be assumed. In writing this book I have assumed that prospective readers will have some familiarity with basic probability and statistics. An understanding of hydraulics (pipe flow, open channel flow and coastal processes) is also required for a full appreciation of the applications and worked examples presented throughout the text and in Chapter 6, although basic background material is provided in each case. The text is aimed primarily at final-year undergraduate and MSc postgraduate students, to bridge the gap between introductory texts and the mainstream literature of academic papers and specialist guidance manuals. I hope it will also be of interest and assistance to practitioners, both those beginning their careers and established professionals requiring an introduction to this rapidly growing discipline.

The motivation for this book arose because, although there are a number of books on reliability theory, these focus predominantly on structural rather than hydraulics problems. Thus, those wishing to use the tools of reliability theory in the context of hydraulic and coastal engineering found a gap between the theory, on the one hand, and on the other hand the practical application of these in their area of interest.

Finally, a few words of caution. While reliability theory and probabilistic methods can help to quantify uncertainty, they should not be thought of as the solution to all problems. Indeed, one of the largest risks in using these methods is that they are applied in a mechanical manner without using all the information that may be available. A good example of this is when there is a discrepancy between the results of a reliability analysis and experience. This can occur when either experience has been incorrectly interpreted or there is an error in one or more elements of the reliability calculations. An investigation of the causes of such a discrepancy will often lead to an improved understanding of the nature of the design problem, and thence to a more satisfactory solution.

The book is divided into seven chapters. A full list of references is given towards the end of the book, and some additional sources of material are cited at the end of individual chapters. The book starts with an introduction to the subject of risk and reliability as it relates to hydraulics and fluid mechanics. Chapters 2 and 3 contain basic and more advanced background material, respectively, on probability, statistics and stochastic processes. In Chapter 4 the subjects of extreme values and the distribution functions of extreme values are covered. The various levels of reliability theory, and their variants, are presented in Chapter 5. Chapter 6 is devoted entirely to worked examples of reliability calculations and problems using the techniques developed in Chapter 5. Problems cover a range of hydraulic and coastal engineering topics. Towards the end of Chapter 6 there are several examples that discuss methods which have been adopted to cope with more 'real-life' cases where measurements or understanding is less complete, and take the reader towards some areas of current research. The book concludes in Chapter 7 with a discussion of the reliability of groups of structures in the context of flood defence schemes.

Acknowledgements

Many colleagues and friends have helped in the writing of this book. I am particularly grateful to Dr Shunqi Pan and Dr David Simmonds for their many useful comments, to Ian Townend who introduced me to reliability theory and to Dr José Horrillo-Caraballo who prepared many of the data and probability plots. I would also like to acknowledge the contributions of Dr Adrián Pedrozo-Acuña and Dr Anna Zacharioudaki (University of Plymouth), Dr Peter Hawkes and Dr Dominic Hames (HR Wallingford), Dr Mark Spivack (DAMTP, University of Cambridge), Dr Grzegorz Różyński (IBWPAN, Polish Academy of Science), Dr Baoxing Wang (ABPMer), Dr Li Ying (DeepSea Research), Prof Yeou-Koung Tung (Hong Kong University of Science and Technology) and the consultants and agencies whose work has provided many of the case studies included in the book.

Principal Notation

A, B	sets; empirical coefficients (wave overtopping)
c	wave celerity (phase speed)
c_g	group velocity
Cov(x, y)	covariance of x and y
$C(x_1, x_2)$	autocovariance
C_{HW}	Hazen–Williams coefficient
C_V, C_C	coefficients of velocity and contraction respectively
C_w	width of permeable crest
$^nC^r$	number of combinations of r objects taken from a set of n objects
d_c	depth of closure
D	pile or pipe diameter, scour depth at a bridge pier
D_{50}	median grain size
D_n	nominal rock diameter
D_{n50}	nominal rock diameter (50th percentile)
E	total wave energy per unit area of ocean, an event
$E(.)$, <.>	the mean or expected value
f	wave frequency
$f_X(x)$	density function of X [$= \Pr(X = x)$]
Fr	Froude number
$F_X(x)$	cumulative distribution function of X [$= \Pr(X \leq x)$]
g	acceleration due to gravity
G	centre of gravity
$G(R, S)$	reliability function
$G(f, \theta)$	directional spreading function
h	water depth
$h(x)$	hazard function
$h(x, y, t)$	seabed levels
H	wave height
$H(x)$	cumulative hazard function
H_b	breaking wave height
H_c	mean height between wave crests

H_{max}	maximum difference between adjacent crest and trough
H_{rms}	root-mean-square wave height
H_s	significant wave height
H_z	mean height between zero upward crossing
$H_{1/3}$	mean height of the highest one-third of the waves
$H_{1/10}$	mean height of the highest one-tenth of the waves
I_R	moment of inertia about the roll axis
k	wave number $(= 2\pi/L)$
k_s	Nikuradse roughness
K	coastal constant for littoral transport
K_D	Hudson's non-dimensional stability factor
L	wave length
$L(\ldots,\ldots,\ldots,\ldots)$	likelihood function
m_n	n^{th} spectral moment, n^{th} ordinary moment of a density function
M	metacentre
$M(s)$	moment generating function
M_n	n^{th} sample moment, n^{th} central moment of a density function
n	Manning's n
N	number of waves during design storm
$N(\mu,\sigma)$	normal distribution with mean μ and standard deviation σ
p	pressure, porosity
P	permeability coefficient
$P(x \geq X)$	probability that random variable x takes on a value greater than or equal to X
$P(A\mid B)$	conditional probability of event A, given that event B has occurred
P_f	probability of failure
$P_{f,n}$	probability of failure over n years
$^nP^r$	number of permutations of r objects taken from a set of n objects
Q	volumetric longshore transport rate
Q_m	mean wave overtopping rate $(m^3/m/s)$
Q_T	discharge rate from an orifice $(m^3/m/s)$
Q_*	dimensionless overtopping rate
r	roughness coefficient
R	strength function in reliability analysis, hydraulic radius
R_*	dimensionless run-up coefficient
R_c	crest level relative to still water level (freeboard)
$R(x_1,x_2)$	autocorrelation function
Re	Reynolds number

S	load function in reliability analysis, channel slope
S_d	damage level parameter
$S(f)$	spectral energy density
$S(f,\theta)$	directional energy density
$S(\omega)$	power spectrum (or spectral density) of a random process
t	time
T	wave period
T_c	mean period between wave crests
T_m	mean wave period
T_p	peak period $(=1/f_p$, where f_p is the frequency at the maximum value of the frequency spectrum)
T_s	significant wave period
T_z	mean period between zero upward (or downward) crossings
u, v, w	components of velocity in the x, y and z directions respectively
U	universal set, upthrust characteristic velocity
$U(a, b)$	uniform distribution between a and b
$\text{Var}(.)$	variance of .
W	sediment size parameter $(=d_{84}/d_{50})$; weight
x, y, z	ordinates in horizontal and vertical directions
X, Y	random variables
z	standard normal variate
Z_0	mean water level above (or below) local datum
α	angle between wave crest and seabed contour
β	reliability index; distribution function parameter
γ	wave breaking index
δ	GEV distribution function parameter
$\varphi(z)$	standard normal density function
$\Phi(z)$	cumulative standard normal distribution function
$\Phi(\omega)$	characteristic function
$\varnothing$	empty set
$\Gamma(x)$	gamma function
$\Gamma_S, \Gamma_R, \Gamma$	safety factors for load, strength and combined effects respectively
η	water surface elevation above a fixed datum
λ	pipe friction factor, distribution function parameter
μ	mean value, distribution function parameter
ν	kinematic viscosity
ρ	density of water, linear correlation coefficient
ρ_r	density of rock

σ standard deviation; distribution function
      parameter
ξ Iribarren number; distribution function
      parameter
ξ_b Iribarren number at wave breaking
ω wave frequency $(= 2\pi/T)$

1 Introduction

1.1 Historical context

Engineering, in some shape or form, has always been a central part of human activity. The construction of shelter from the elements, whether it be a single-storey shack or a flat in a high-rise tower block; the means for transporting goods and people by road, rail, ship and air; the provision of drinking water and hygienic drainage; all require careful design if they are to be effective and economically feasible. It may come as a surprise to realise that the coastline of many countries has long been as much engineered as natural. For example, the Port of A-ur was built on the Nile prior to 3000 BC. Nearby on the open coast, the Port of Pharos was constructed around 2000 BC, and had a massive breakwater which was over 2.5 km long. The Romans developed the practice of pile driving for cofferdam foundations, a technique that was used for the construction of concrete sea walls. Much more recently, the Dutch reclaimed a large area of land from the sea in the form of polders. In the 1950s, a 70 km long embankment was constructed through the sea, turning the South Sea or *Zuiderzee* into a lake now known as the *Ijsselmeer*. In the very recent past, our understanding of materials and physical processes has extended the concept of engineering to the design and construction of offshore archipelagos such as those in Dubai, as shown in Figure 1.1.

Whilst early structures were no doubt built on the basis of trial and error, this would have been informed by observing what led to both successes and failures. With Isaac Newton's development of mechanics, the path to a rational and quantitative approach to structural design was set. Disciplines involving the flow of fluids had to wait a bit longer. The Swiss mathematician Leonhard Euler was invited by Frederick the Great to Potsdam in 1741. By popular account he was asked to make a water fountain. As a theoretician, he began by trying to understand the laws of motion of fluids, using Newton's laws for a fluid. This equation predicts velocities that are much higher than anything observed. In fact, failure was inevitable because the equations used by Euler omitted important processes, one of which was viscous dissipation, that is, the dissipation of energy due to friction.

Figure 1.1 Man-made archipelago in Dubai, United Arab Emirates. Groups of
islands in the shape of palm trees have been constructed by building up
the seabed with material dredged from elsewhere. The palm islands are
called Palm Jebel Ali, Palm Jumeirah and Palm Deira, respectively, from
the south to the north. The largest island, Palm Deira, was just being
started (towards the top right) when this image was taken. Between Palm
Jumeirah and Palm Deira there is 'The World', a cluster of 250–300 small
islands in the shape of a world map. The ASTER image was acquired on
18 September 2006, is centred near 25.3°N, 55.3°E, and covers an area
of 46.1 × 57 km. This image is reproduced with acknowledgement to
the NASA/GSFC/METI/ERSDAC/JAROS, and US/Japan ASTER Science
Team.

The appropriate term was added by engineer Claude-Louis Navier in 1827
and mathematician Gabriel Stokes in 1845 to develop a theoretical framework for Newtonian fluids. The resulting equations are known as the
'Navier-Stokes equations'. For flow in a single dimension, the equation may
be written for unit fluid volume as:

$$\frac{\partial u}{\partial t} + u\frac{\partial u}{\partial x} = -\frac{1}{\rho}\frac{\partial p}{\partial x} + \nu\frac{\partial^2 u}{\partial x^2} \tag{1.1}$$

where u is the fluid velocity, t is time, x is the independent direction coordinate, p is pressure, ρ is the density of the fluid and v is known as the kinematic viscosity. The detailed derivation of this equation is beyond the scope of this book. However, it will be helpful to understand the physical principles behind the terms in the equation. The terms on the left-hand side of the equation describe the change in the fluid velocity following the motion of the fluid. If the first term is positive, the fluid velocity at a fixed point will be increasing with time. If the second term is positive, the fluid is accelerating in the direction of increasing x, at a fixed time. Figure 1.2 shows two cases where the terms on the left-hand side are positive.

The first term on the right-hand side arises from the force acting on the fluid due to differences of pressure in space. The negative sign is required because fluid will tend to flow from regions of high pressure to regions where the pressure is lower, as illustrated in Figure 1.3. The force due to the Earth's gravity is often implicitly included in this term. The second term on the right-hand side of Equation (1.1) represents the effects of friction. This is diffusive in nature and has the effect of 'smoothing' small-scale spatial variations in velocity. Without this term, kinetic energy is conserved; whereas, when this term is included, kinetic energy is dissipated.

Simplistic attempts to use Equation (1.1) can lead to very unrealistic conclusions. For example, consider the flow of water down a river to the sea. For fresh water, typical values of density and kinematic viscosity are $1000\,kg/m^3$ and $1.5 \times 10^{-5}\,m^2/s$ respectively. A river such as the Nile drops hundreds of metres over the course of approximately 100 km. The typical slope of the river is therefore $\tan(10^{-4}) \approx 10^{-4}$. Assuming that there is a balance between the gravitational force and the viscous dissipation leads to $u \approx 10^5\,m/s$, instead of the observed value of approximately 1 m/s. This result is, of course, ridiculous and points to the inappropriateness of this assumption. Alternatively, arguing on the lines of energy conservation then, at best, all the potential energy lost by the fluid in moving from the head to the base of the river is converted to kinetic energy. The stored potential energy is $\rho g H$, where H is the change in elevation. In the case of the Nile

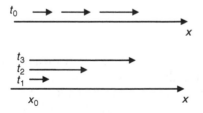

Figure 1.2 Top – acceleration as a velocity gradient in space at a single instant in time; Bottom – acceleration as a velocity gradient in time at a fixed location.

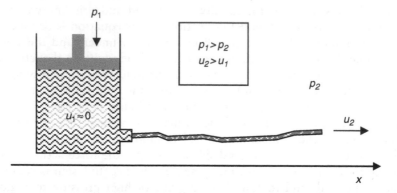

Figure 1.3 Fluid virtually at rest in a tank is put under pressure, p_1, so that it flows out along a tube open to atmospheric pressure p_2. There is a pressure difference (gradient) in the fluid between the tank and at the end of the tube. The pressure decreases in the direction of increasing x, and therefore the gradient is negative. The corresponding gradient in the fluid velocity is positive, as fluid in the tank is virtually at rest but exits the tube at a large positive velocity v_2.

$H \approx 500$ m and the potential energy loss is $\approx 5 \times 10^6$ J. Equating this with the kinetic energy gain gives $u \approx \sqrt{(2gH)} \approx 100$ m/s. This is a step in the right direction, but is still unrealistic.

In 1894, Osborne Reynolds suggested a resolution of these discrepancies. Reynolds performed a series of experiments to investigate the relative importance of the terms in Equation (1.1). In particular, he noted that the nature of the flow depended on the relative magnitude of the viscous term (second term on the right-hand side of Equation 1.1) and the nonlinear term (second term on the left-hand side of Equation 1.1). The ratio of the nonlinear term to the viscous term can be measured by the quantity UL/v, where U is a characteristic value of the velocity and L is a characteristic length scale such as river width or pipe diameter. This quantity is now known as the Reynolds Number (Re):

$$Re = \frac{UL}{v} \tag{1.2}$$

and plays an important role in hydraulic engineering. If $Re \ll 1$ then the nonlinear term can be neglected, simplifying the solution of the Navier-Stokes equations considerably. However, in many natural situations $Re \gg 1$ and no such simplification is possible. The flows are strongly affected by nonlinearity, and the flow patterns can be highly complicated and distorted. Such flows are termed 'turbulent'. For example, in the case of the river flow discussed above, the Reynolds number is $\sim 10^7$, and the flow may be expected to be turbulent. Turbulence generally has a damping effect on the mean flow, and is the cause of the discrepancy between the river flow velocities calculated on simple physical arguments and those observed. It is also likely to have been a contributory cause to Euler's problem in designing a fountain.

The examples above serve to illustrate how the process of engineering is an evolutionary one. Existing understanding is employed in a new design. If the design works, all well and good; if it does not, then through the sequence of testing and further scientific research an understanding of why the design did not work can emerge which, in turn, can lead to an improved design.

1.2 Uncertainty

Hydraulic and coastal engineering design deals with the flow of water and its effects on the built and natural environment. This involves not just the physical processes but also an appreciation of the wider interactions with the environment in terms of ecology, archaeology and socio-economics. In this regard it is worth considering that any engineering works are initiated as a result of a human need. For example, this might be for housing, for a bridge to enable faster communication between communities, for protection against flooding, for national esteem or to satisfy personal vanity. In all cases, the engineering works will have an associated cost in terms of money to pay for the materials and construction, as well as perhaps less tangible side-effects. In the case of a flood defence this might be restricted access to a river for tourists, which constrains scope for touristic development of the site. The funding of an engineering scheme will inevitably involve an assessment of the balance between the costs of the scheme, on the one hand, and the value and benefits it would bring, on the other. Clearly, it would not be sensible to proceed with a scheme the costs of which exceeded the expected benefits. However, the way in which benefits are costed, particularly when human life is involved, is a difficult and emotive subject.

An additional complication is that all materials deteriorate over time, reducing the integrity of any structure or engineering scheme. This is acknowledged in current engineering design philosophy as follows. For the sake of argument, let us consider the construction of an embankment as a flood defence. Once built, this will be subject to the vagaries of the weather, the river flows, progressive consolidation of the embankment material and possibly subsidence. With some knowledge of the processes of erosion and consolidation, the rate at which the embankment crest level is likely to reduce can be estimated and thus included in the original design. In order to do this, however, it is required to specify a duration over which the structure is expected to provide protection. This is known as the 'design life'.

As well as the design life, a designer also needs to know what type of extreme condition event the structure is required to resist. This is often expressed in terms of a 'level of service', or in other words a statement such as 'prevent flooding against all water levels up to and including that experienced once every 50 years'. This implicitly acknowledges the fact that, however high you build the embankment, during the course of the design life there *may* be a very large flood that overwhelms the defence. For example, if water levels of a magnitude that were experienced only once every 100 years occurred, you could not reasonably expect a defence that had been designed to protect against the once in 50 year levels to prevent flooding. Of course you could decide to improve the design to protect against the 1 in 100 year level. However, this would be more expensive, reducing the benefit–cost ratio, and would still not guarantee protection against even more severe conditions.

From this perspective it becomes clear that the design process is a balance between the level of protection required, the expected lifetime of the defence, the costs of construction and the value of the benefits. The costs of construction are determined by purely financial pressures, the benefits largely by environmental and socio-economic considerations, the design life and level of service by *society's attitude to the consequences of the failure of the structure*. This latter factor varies from discipline to discipline and from country to country. An example of this is that, in the UK, sewers are typically designed to provide a level of service that equates to the 1 in 20 year flow in urban areas over a design life of 30 years (although this is an area of active discussion – see e.g., http://www.fwr.org/wapug/wapugdes.pdf), while river and coastal embankments are designed to a level of 1 in 50 years or more over a design life of 50 years. In contrast, coastal defences in Holland are designed to protect against the 1 in 10,000 year storm as about 50% of the land area of Holland is protected by sea and river flood defences.

It might seem that once a design life and a level of service have been specified there should be few problems. However, due to our imperfect knowledge of the physics of fluid flow, it is not possible to be absolutely certain how a structure will perform. Unpredictability also arises from the nature of the materials used in construction. Soils and sands are composed of an aggregation of individual grains that have varying density, volume, shape and orientation. The voids between the grains are similarly of different shapes and volumes, and may permit the flow of fluid through the soil. Quarried stone and concrete contain microcrystalline imperfections or faults (Radavich 1980), which permit the formation and propagation of cracks. Indeed, it is impossible to quarry two stones or manufacture two concrete blocks that are identical at a *microscopic level*. This fact means that there is some uncertainty about how any one of, say, 100 concrete blocks will actually perform.

Uncertainties also arise not just from our imperfect understanding of environmental processes and materials, but also in the implementation of the design (e.g., using substitute materials that have different properties, slipshod construction, reduction/redesign of the level of service during construction due to economic constraints, and so on). A large uncertainty will often be compensated, through experience, by an appropriate allowance for contingency costs and conservatism in the design. Reducing uncertainty can translate into more cost-efficient engineering solutions. Engineering is an inherently uncertain and iterative process, with design guidelines changing according to improvements in our understanding, the introduction of new materials and our attitude to risk.

1.3 Risk and Reliability

1.3.1 Introduction

We all, consciously or unconsciously, assess risks every day of our lives – at work, at leisure and in the home. In general, individuals seem to be prepared to take on greater risks than are commonly permitted for public services and transport. This is a reflection of society's consensus that a low level of risk should be provided to individual members of the public when they entrust their safety to others. Individuals who choose to engage in more dangerous activities may do so, but at their own risk.

Table 1.1 summarises the risk of various activities by the rate of fatalities. Not surprisingly, activities such as skydiving and motorcycling are high risk in these terms. However, this is only one way of measuring risk, and is perhaps not the fairest, because even a successful skydiver is unlikely to spend as much time skydiving as a passenger would spend travelling on buses or long-haul flights. An alternative measure would be to calculate the number of fatalities occurring in a unit of time. Table 1.2 presents estimated figures in this format. Firstly, not all the same activities appear, as some activities are easier to measure than others.

Secondly, there are some surprising figures. The fact that you are reading this book means that, according to Table 1.2, you have already succeeded in surviving the very dangerous process of birth. Also, it appears that climbing stairs is more dangerous than amateur boxing. Furthermore, travelling by bus or train is much more dangerous in the UK than in the USA.

Clearly, a simple measure of risk, such as those in Table 1.1 or 1.2, is open to misinterpretation. For example, there may be a simple reason for the discrepancies in the risk of travelling by bus or train in the USA and UK, such as different train speeds, so that in a fixed time trains cover more distance in one case than in the other, or differences in the passenger density on trains in the two cases. Nevertheless, when taken in context, such tables provide the means of a quantitative comparison of the risks associated with various activities.

1.3.2 What is risk?

As suggested by the previous section, the word 'risk' implies a measure of the likelihood of an event occurring over a particular period of time. In slightly more formal language,

$$\text{Risk} = \text{hazard} \times \text{consequence} \tag{1.3}$$

where 'hazard' is the likelihood of an event happening in a given time, and 'consequence' is a measure of the damage *given that the hazard event occurs*. The hazard will usually be expressed in terms of a probability of occurrence per unit time, and the consequences measured in fatalities or financial costs. A formal definition of risk is given in Chapter 2, after the introduction of the notion of probability.

1.3.3   What is reliability?

To define reliability we return to the idea of a structure designed to withstand particular conditions over the design life. Unreliable structures are often much easier to identify than reliable ones, as failure often captures the headlines. However, 'failure' of a structure is not always straightforward to assess. If a building collapses or a dam breaches then it is failure. But what if a flood embankment designed to withstand a 1 in 50 year flood is overtopped but negligible damage occurs? How much and what kind of damage must occur before flooding is considered substantive? Was the event a 1 in 50 year or a 1 in 51 year event? Do we have sufficient data and understanding to measure extreme events to this resolution? While determining failure can be very subjective and qualitative, not to mention emotive and litigious,

Table 1.1 Comparative risk of death for a range of activities (adapted from Failure Analysis Associates, Inc., 1993)

Activity	Fatalities per million hours
Skydiving	128.71
General aviation	15.58
On-road motorcycling	8.80
Scuba diving	1.98
Swimming	1.07
Motor cars	0.47
Water-skiing	0.28
Cycling	0.26
Flying (scheduled domestic airlines)	0.15
Cosmic radiation during long-haul flights	0.035
Home living (active)	0.027
Travelling in a school bus	0.022
Home living, active and passive (sleeping)	0.014
Residential fire	0.003

Table 1.2 Comparative risk of death for a range of activities, measured in fatalities per billion per hour of the activity (adapted from http://zebu.uoregon.edu/1999/ph161/l20.html) (accessed 24/01/08)

Activity	Deaths per billion with 1 h of risk exposure
Birth	80,000
Professional boxing	70,000
On-road motorcycling	6,280
Swimming	3,650
Cigarette smoking	2,600
Flying (scheduled airlines)	1,450
Motor cars	1,200
Coal mining	910
Climbing stairs	550
Amateur boxing	450
Travelling in a bus or train (UK)	50
Travelling in a bus or train (USA)	10

its opposite, the measure of success or 'reliability', can often be defined and quantified in an objective manner.

One widespread definition of reliability is: 'Reliability is the probability of the structure (or system) performing its required function adequately for a specified period of time under stated conditions'. This definition is based on the concept of level of service, and acknowledges that responsibility falls on the commissioning body and designer to agree what is to be meant by the 'required function', 'adequately', 'design life' and 'design conditions'. A key element of the definition of reliability is the word 'probability'. The forces experienced by engineering structures are inherently random (e.g., wind, rain, waves) and the response of the materials to these forces are uncertain. Describing both the forces and responses in probabilistic terms provides the basis for objective and quantitative description of the uncertainties in structural performance. Furthermore, the natural language for bringing the ideas of uncertainty, risk and reliability together is probability.

The study of risk and reliability of structures, often termed 'reliability theory', has been developed widely in structural engineering and manufacturing. Hence, much of the nomenclature is cast in the language of these disciplines. Thus, a structure is considered to have a certain 'strength' to resist the imposed 'loads', whether these are wind stresses acting on a tall building or water levels approaching the crest of an embankment.

Reliability theory provides a means of assessing the performance of existing structures and quantifying the uncertainties associated with new ones. Reliability theory can also be very useful in determining the relative risks of failure of a collection of existing structures. For a body responsible for maintenance of structures such as dams or flood defences, this can assist in prioritising limited funds to the weakest structures. A final caveat; although reliability theory allows us to determine the likelihood of failure, it is important to remember that the calculations of probabilities have themselves an associated uncertainty due to the limited accuracy of data and the approximate nature of many of the equations defining the response of the structure to loads. The results of reliability calculations should always be used within the context of engineering experience.

1.4 Scope

Reliability theory is well established in civil engineering in the discipline of structural engineering. Its uptake into hydraulic and coastal engineering is more recent, and is leading to new developments of the theory to deal with the unique problems that arise when flowing water is involved. The theory requires knowledge in a number of specialist subjects, including hydraulics, wave mechanics, sediment transport, probability and numerical methods.

An appreciation of the power and limitations of numerical prediction methods is becoming increasingly important due to the improvements in computing power and the development of computational methods for describing hydrodynamics and uncertainty. Indeed, some aspects of reliability analysis, such as Monte Carlo simulation, can be effective only with numerical methods. The use of some numerical techniques is covered in this book, and references to more detailed methods are provided where appropriate.

Within coastal engineering, and also hydraulic engineering, there is a clear trend towards 'soft engineering'. Soft engineering does not exclude hard structures, but promotes strategic design that takes into account the impact that construction will have on the surrounding environment. In turn, this requires a much more detailed appreciation of the natural processes in design, as well as a recognition of those areas where our understanding is lacking. This generally means the use of complex numerical models for predicting fluid flow, sediment transport and so on, all of which introduce further uncertainty. Some such models are described in this book but, for clarity, well-known empirical design formulae are used for many of the examples.

This book is intended to provide an introduction to reliability theory in its application to hydraulic and coastal engineering. It includes elements of probability theory and the analysis of extremes prior to the description of reliability theory in Chapter 5. Much design and assessment work now makes use of mathematical or numerical models, and the use of such models is a continuing thread throughout the book. Chapter 6 is devoted to the application of reliability methods in a range of contexts. At the beginning of each subsection in Chapter 6, a brief discussion of the necessary theory and/or empirical design equations are provided. Chapter 7 provides an introduction to how the methods of reliability theory described in the earlier chapters can be applied to systems containing many linked elements, and the concept of system reliability.

A key part of the engineer's repertoire is judgement, based on experience gained from design and construction projects. While it is not possible to teach design experience, the examples included in this book have been selected to expose the reader to a spectrum of problems; from those that are well defined with good-quality data, to those where the problem is not so well defined or the data are imperfect. It is hoped that these convey the element of pragmatism that is often required in practical design problems, in which one rarely has as much information as one would like, or indeed needs, to undertake detailed calculations. This is not to say that calculations are not necessary; a good understanding of basic statistical and reliability theory is important in making informed and robust judgements.

Further reading

Batchelor, G. K., 1967. *An Introduction to Fluid Dynamics*, Cambridge University Press, London.

Chadwick, A. J., Morfett, J. and Borthwick, M., 2007. *Hydraulics in Civil and Environmental Engineering*, 4th edition, E. & F. N. Spon, Abingdon.

Melchers, R. E., 1999. *Structural Reliability Analysis and Prediction*, 2nd edition, John Wiley & Sons, Chichester, p. 437.

Nakayama, Y. and Boucher, R. F., 1999. *Introduction to Fluid Mechanics*, Arnold, London.

Reeve, D. E., Chadwick, A. J. and Fleming, C. A., 2004. *Coastal Engineering: Processes, Theory and Design Practice*, SPON Press, Abingdon, p. 461.

Roberson, J. A., Cassidy, J. J. and Chaudry, M. H., 1998. *Hydraulic Engineering*, 2nd edition, John Wiley & Sons, New York.

Thoft-Christensen, P. and Baker, M. J., 1982. *Structural Reliability Theory and Its Applications*, Springer, Berlin.

Versteeg, H. K. and Malalasekera, W., 1995. *An Introduction to Computational Fluid Dynamics*, Longman, Harlow.

2 Introduction to probability

2.1 Probability

Many statements made in everyday conversation contain descriptions of uncertainty: 'It may rain tomorrow'; 'That embankment should be high enough to hold against this year's storms'; 'I wouldn't do that if I were you'. The theory of probability provides a basis for constructing exact measures of uncertainty. The theory of probability is often considered to have begun in the mid-sixteenth century, when the Italian mathematician and gambler Girolamo Cardano wrote his 'book on games of chance'. Many of the ideas of probability theory can be illustrated through examples relating to dice and cards, and some such examples are used in what follows. In this chapter, a brief introduction is given to some of the concepts underlying reliability theory. The treatment is not intended to be exhaustive, and readers seeking further detail are referred to the list of further reading at the end of the chapter.

2.1.1 Sets and relations

In order to understand probability, we need the concept of a set. A set is a collection of distinguishable objects, for example: the letters a, b, c, x, y and z; the numerals 1, 2, 3 and 4; the primary colours red, blue and yellow. A set is usually denoted by a pair of braces, { }, enclosing the elements of the set. Thus, we might write that A denotes the set $\{1, 2, 3, 4\}$, or $A = \{1, 2, 3, 4\}$. Note that the order of the elements does not matter.

Two sets A and B are equal if every element of one is an element of the other. A set A is *a subset* of a set B if every element of A is also an element of B. The set that has no elements, the *empty set*, is denoted by $\emptyset$, and is a subset of all sets. The union of two sets is the set of all elements that belong to A or B or both A and B. The union of sets A and B is written as $A \cup B$. The intersection of two sets A and B is the set containing elements that are in both A and B, and is written as $A \cap B$. Two sets A and B are said to be *disjoint* or *mutually exclusive* if they have no elements in common, that is, if $A \cap B = \emptyset$.

Example 2.1. If $A = \{2,3,4\}$ and $B = \{3,4,5,6\}$ then $A \cup B = \{2,3,4,5,6\}$ and $A \cap B = \{3,4\}$

Example 2.2. If $A = \{2,3,4\}$ and $B = \{5,6\}$ then $A \cup B = \{2,3,4,5,6\}$ and $A \cap B = \emptyset$

In any particular problem or application there will normally be a known range of outcomes. For example, when rolling a fair die one would expect to roll one of the integers 1, 2, 3, 4, 5 or 6. The set of all outcomes for a particular situation is termed the *universal set*, or U. In the case of a die, $U = \{1,2,3,4,5,6\}$. Referring to this set, we may be interested in different subsets of U, such as the even integers, or those integers greater than 4. If A and B are two *disjoint* sets such that $A \cup B = U$, they are said to be *complementary* (with respect to U). The *complement* of a set A, that is, the set containing all elements in the universal set not in A, is often written as A'.

Example 2.3. If $U = \{1,2,3,4,5,6\}$ and $A = \{1,3,5\}$ then $A' = \{2,4,6\}$. Note that $A \cap A' = \emptyset$ and $A \cup A' = U$.

A useful pictorial means of understanding set theory is the Venn diagram. In this, the elements of a set are represented by points and the set itself by an aggregate of the points contained within a circle. The universal set is usually shown as a rectangle. In Figure 2.1 the overlapping section of the circles A and B represent two intersecting subsets of U. In Figure 2.2 A and B represent two disjoint sets.

Example 2.4. Suppose that along a river reach there are 50 flood defence units which are either embankments or vertical walls. Of these, 14 are walls, 27 have crest protection and 8 are walls and have no crest protection. How many embankments do not have crest protection?

Solution. Writing W for the set of defences that are walls, C for the set of defences that have crest protection, then we wish to know how many defences are in $W' \cap C'$. The steps in the solution are:

Number of walls without crest protection $= W \cap C'$	8
Walls with crest protection $= W \cap C$	$14 - 8 = 6$
Embankments with crest protection $= W' \cap C$	$27 - 6 = 21$
Embankments without crest protection $= W' \cap C'$	$50 - (21 + 6 + 8) = 15$

This simple problem may also be solved with the aid of a Venn diagram (Figure 2.3).

While Venn diagrams are a useful tool when the number of sets is relatively small, they become cumbersome with large numbers of sets. To deal

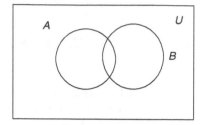

Figure 2.1 Intersecting sets.

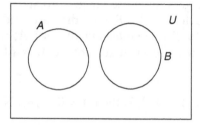

Figure 2.2 Disjoint sets.

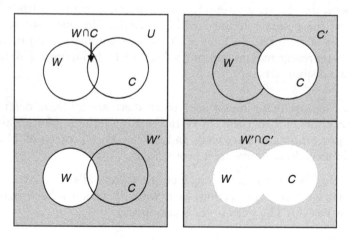

Figure 2.3 Venn diagram for Example 2.4. Top left: the two intersecting sets W and C in the universal set U. Top right: the set C' (shaded). Lower left: the set W' (shaded). Bottom right: the intersection of W' and C' (shaded).

with this, it is necessary to define rules (or axioms), that define how we can manipulate sets. The following eight 'laws' define the rules governing sets:

1   $A \cap B = B \cap A$ and $A \cup B = B \cup A$
2   $A \cap (B \cap C) = (A \cap B) \cap C$

3   $A \cup (B \cup C) = (A \cup B) \cup C$
4   $A \cap (B \cup C) = (A \cap B) \cup (A \cap C)$
5   $A \cup (B \cap C) = (A \cup B) \cap (A \cup C)$
6   $A \cap \emptyset = \emptyset$ and $A \cup \emptyset = A$
7   $A \cap U = A$ and $A \cup U = U$
8   $A \cap A' = \emptyset$ and $A \cup A' = U$

where A, B and C are sets, U is the universal set and $\emptyset$ is the empty set. Two relations proved by the English mathematician De Morgan will also be helpful later. They are:

9   $(A \cup B)' = A' \cap B'$
10   $(A \cap B)' = A' \cup B'$

It is left as an exercise for the reader to convince themselves that *De Morgan's rules* are true. (*Hint.* Use a Venn diagram with intersecting sets A and B, inside the universal set U.)

2.1.2 Factorials

For any positive integer n, we define n *factorial*, denoted by $n!$, as follows:

$$n! = n(n-1)(n-2)\ldots3.2.1 \tag{2.1}$$

Thus $3! = 3.2.1 = 6$ and $8! = 8.7.6.5.4.3.2.1 = 40320$.

2.1.3 Permutations and combinations

Let $A = \{x, y, z\}$. A *permutation* of this set is any ordered group of elements, that is, any subset of the set in which the order of the elements is noted. An unordered group of elements is termed a *combination*. The ordered triples of set A are: (x, y, z), (x, z, y), (y, x, z), (y, z, x), (z, x, y) and (z, y, x). Each of the six groups contains the same combination of elements, but arranged in a different order. If we now select two letters from the three in set A we can make six ordered pairs or permutations: (x, y), (y, x), (x, z), (z, x), (y, z) and (z, y). But (x, y) and (y, x) are the same combination of elements, and thus, three combinations are possible, if we take two elements at a time. We will use two important results regarding permutations and combinations. These are now given without proof.

The number of distinct permutations of r objects taken from a set of n objects is:

$$^nP^r = n!/(n-r)! \tag{2.2}$$

The number of combinations of r objects taken from a set of n objects is:

$$^nC^r = n!/[r!(n-r)!] \tag{2.3}$$

Example 2.5. A design competition awards prizes for 1st, 2nd and 3rd places. If five engineers enter the competition, in how many ways may the prizes be won?

Solution. This is a question about permutations, where $n = 5$ and $r = 3$. Hence, by Equation (2.2), $^5P^3 = 5!/2! = 60$.

Example 2.6. A committee of 6 is to be chosen from a group of 14 people. In how many ways may the committee be chosen?

Solution. This is a combination question, where $n = 14$ and $r = 6$. Hence, from Equation (2.3), $^{14}C^6 = 14!/(6!8!) = 14.13.12.11.10.9/6.5.4.3.2.1 = 3003$.

2.1.4 *Sample space*

The set of all possible outcomes of an experiment is termed the *sample space*, often denoted by the letter S. Here, the result of an 'experiment' is called an 'outcome'. An 'experiment' may well be an experiment in the everyday sense of the word, such as applying a load to a concrete beam. The outcomes of such an experiment would be either that the beam supports the load successfully or that it does not. However, the term 'experiment' may also be applied to rather different situations, such as finding the annual maximum water level at a tide gauge. In this case, the outcome of the experiment (the annual maximum water level) is not constrained to take on discrete values such as 1, 2 or 3, but instead takes one from a continuum of values. We define an *event* as a subset of the sample space. Events may be outcomes, but do not necessarily have to be, as the following example demonstrates.

Example 2.7. A die is thrown. The outcomes are $\{1, 2, 3, 4, 5, 6\}$. The following are events:

  I The number 5 is thrown;
 II An even number is thrown;
III A number greater than 3 is thrown.

In I the event is one of the outcomes. In II and III the event occurs if any one of three outcomes occurs: $\{2, 4, 6\}$ and $\{4, 5, 6\}$, respectively.

Two events E_1 and E_2 are called *mutually exclusive* if they are disjoint, i.e., if $E_1 \cap E_2 = \emptyset$.

2.1.5   *Random variables and probability*

A *random variable* X is a function defined over a sample space $S = \{a_1, a_2, a_3, \ldots, a_n\}$. This means that X takes on a unique value for each element a_i of S. The outcome a_i determines a single value of the function, which we denote by x_i. An example may help to clarify this rather formal definition.

Example 2.8. Two dice are thrown. The sample space consists of 36 ordered pairs: $\{(1, 1), (1, 2), \ldots, (1, 6), (2, 1), (2, 2), \ldots, (6, 1), (6, 2), \ldots, (6,6)\}$. Now define a random variable X as the sum of the numbers appearing on the two dice. Then: $X(1, 1) = 2$, $X(1, 2) = X(2, 1) = 3$, $X(1, 3) = X(2, 2) = X(3, 1) = 4$, and so on.

Note that each outcome determines a single value of the random variable X, but a given value of X may correspond to more than one outcome.

In Example 2.8 it is clear that, although (with fair dice) each of the outcomes would be expected to be equally likely, the same does not apply to the values of the random variable X. Thus, we can throw '2' $= 1 + 1$ in only one way, but we can get a '3' $= 1 + 2 = 2 + 1$ in two ways. Thus, purely from a consideration of the number of ways in which the random variable can take on the same value, we might expect that in a long sequence of throws of two dice that X would take the value 3 twice as many times as it takes the value 2. Slightly more formally, the probability of an event is the number of times the event occurs in N repeated experiments divided by the number of experiments. As N increases in magnitude, our experience of most scientific experimentation indicates that this ratio tends to a constant value. In more abstract terms, one may think of the process of assigning a probability to an event as a function defining a non-negative value to that event in the sample space.

Example 2.9. A hand of 5 cards is dealt from a well-shuffled, standard 52-card deck. What is the probability that it contains 3 spades and 2 clubs?

Solution. The number of possible hands (combinations) of 5 cards is $^{52}C^5 = 52!/(5!47!) = 2,598,960$. Any combination of 5 cards is as likely to be dealt as any other, so the probability of any particular 5-card hand is $1/2,598,960$. Now, the number of ways in which a combination of 3 spades can arise from 13 spades (the total number of spades in the deck) is $^{13}C^3 = 13!/(3!10!) = 286$. Similarly, for the clubs we have $^{13}C^2 = 13!/(2!11!) = 78$. The event, 'three spades and two clubs' can occur in $^{13}C^3 \times {}^{13}C^2$ ways. The required probability is thus $^{13}C^3 \times {}^{13}C^2/2,598,960 = 0.00858$, or approximately once in every 100 deals.

2.1.6   *Axioms of probability*

By convention, there are certain desirable properties for probabilities to have, termed 'axioms'. Writing S as the sample space and $P(E)$ as the probability of an event E occurring in an experiment we define the following characteristics for probabilities:

1   For any event $E, 0 \le P(E) \le 1$
2   If E_i are mutually exclusive events then

$$P(E_1 \cup E_2 \cup \ldots \cup E_n) \equiv P\left(\bigcup_{i=1}^{n} E_i\right) = \sum_{i=1}^{n} P(E_i) \tag{2.4}$$

3   If $E = \varnothing$, then $P(E) = 0$, and the event is said to be impossible
4   If $E = S$, then $P(S) = 1$, and the event E is said to be certain

Equation (2.4) is often known as the *theorem of total probability*. By considering the two events E and its complement E', we know that $E \cup E' = S$. So $P(E \cup E') = 1$. But $P(E \cup E') = P(E) + P(E')$ because E and E' are mutually exclusive. Thus,

$$P(E') = 1 - P(E) \tag{2.5}$$

Equations (2.4) and (2.5) are very important as they will be used repeatedly in later chapters.

2.2   Elementary results from probability theory

2.2.1   *Conditional probability*

Consider the experiment of throwing two dice. For the sake of argument, let us suppose one die is red and the other blue. The outcomes of the experiment can be represented as a square array of 36 points, as shown in Figure 2.4.

What if we now ask the question, 'If the pair of dice show an odd sum, what is the probability that this sum is greater than 7?' This question is rather more complicated than those we have come across so far. It can be broken into two parts. First, we are told that the dice show an odd sum. This automatically restricts the number of outcomes of interest to only 18 out of the total of 36 possible outcomes. Then, of these outcomes we are asked which have a sum greater than 7. There are 6 points out of the 18 which satisfy this criterion, {(6, 3), (5, 4), (4, 5), (3, 6), (6, 5) and (5, 6)}, each having an equal probability. Thus, the answer to the question is that the probability is $6/18 = 1/3$. This is an example of what is termed *conditional probability*. The probability of an event B occurring, given that event

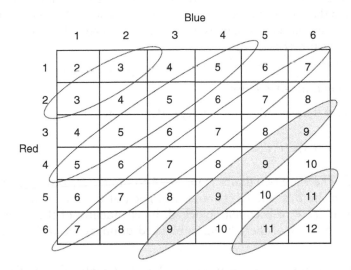

Figure 2.4 The possible outcomes of throwing two dice are shown in a tabular format. The odd sums are circumscribed with a dotted line. The odd sums
that are greater than 7 are, in addition, shaded.

A has already occurred, is denoted $P(B|A)$. The conditional probability is
defined as

$$P(B|A) = \frac{P(A \cap B)}{P(A)} \tag{2.6}$$

The events A and B are called *statistically independent* if

$$P(A \cap B) = P(A)P(B) \tag{2.7}$$

Example 2.10. A coin is tossed twice. What is the probability that both
tosses are 'heads', given that at least one of the tosses is a head?

Solution. There are four possible outcomes: {HH, HT, TH, TT}, where H
denotes 'heads' and T denotes 'tails'. Each outcome has equal probability $=$
$1/4$. Thus,

$$
\begin{aligned}
P(HH|one\ H\ at\ least) &= P(HH|HT \cup TH \cup HH) \\
&= \frac{P(HH \cap (HT \cup TH \cup HH))}{P(HT \cup TH \cup HH)} \\
&= \frac{P(HH)}{P(HT \cup TH \cup HH)} \\
&= \frac{1}{3}
\end{aligned}
$$

Many people are surprised by this answer, thinking that it should be 1/2. However, that is the answer to a similar but distinct question: 'When a coin is tossed twice what is the probability that both are heads given that the second toss is a head?' (*Exercise*).

2.2.2   Conditional independence

The events A and B are conditionally independent given C if

$$P(A \cap B|C) = P(A|C)P(B|C) \qquad (2.8)$$

Example 2.11. There are two drains from site A to site B, (D_1 and D_2), and two drains from site B to site C, (D_3 and D_4). Each of the four drains has an equal probability P of becoming blocked, independently of the others. What is the probability of there being unblocked drainage from site A to site C?

Solution.   The situation is shown in Figure 2.5. The probability of there being an unblocked drain from site A to site C is equal to the probability of there being an unblocked drain from site A to site B and there being an unblocked drain from site B to site C. As the blocking of each of the drains is independent, this is simply the product of the probabilities of there being unblocked drains from A to B and from B to C. However, blockages preventing drainage from A to C can occur in several different ways, {D_1 and D_2 blocked, and neither or only one of D_3 and D_4 blocked; D_3 and D_4 blocked, and neither or one of D_1 and D_2 blocked; all four drains blocked}. Rather than count these and evaluate probabilities for each case we can use Equation (2.5). Now,

$$
\begin{aligned}
P(\text{unblocked} &= 1 - P(\text{blocked drain from } A \text{ to } C) \\
\text{drainage}) \\
&= [1 - P(\text{drains from } A \text{ to } B \text{ blocked})] \times \\
&\quad [1 - P(\text{drains from } B \text{ to } C \text{ blocked})] \\
&= [1 - P(\text{drains from } A \text{ to } B \text{ blocked})]^2 \\
&= [1 - P(\text{drain } D_1 \text{ blocked and drain } D_2 \text{ blocked})]^2 \\
&= [1 - P(\text{drain } D_1 \text{ blocked}) \times P(\text{drain } D_2 \text{ blocked})]^2 \\
&= (1 - p^2)^2
\end{aligned}
$$

As the probabilities of each drain becoming blocked are equal, the probability that the drains from A to B are blocked is the same as the probability that the drains from B to C are blocked. Thus, we can substitute expressions for the first whenever the second occurs – as in going from line 2 to line 3 in the above solution. As the blocking of each of the drains is independent, the probability of both D_1 and D_2 becoming blocked is simply the product of the probabilities, according to Equation (2.7). If we take $p = 1/10$ in the above, the probability of

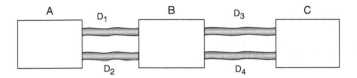

Figure 2.5 Diagram for Example 2.11.

unblocked drainage from A to C is 0.98. Suppose now that drains D_2 and D_4 were not installed. Then the probability of unblocked drainage from A to $C = [1 - P(D_1 \text{ blocked})] \cdot [1 - P(D_3 \text{ blocked})] = (1 - p)^2 = 0.81$. The benefit of having some redundancy in the drainage system (an increase in the probability of functioning drainage of 0.17 in this case) is clear, but in practice would have to be set against the cost of duplicating the drainage routes.

2.2.3 Partitions and Bayes' rule

Consider a sample space S. If this is divided into mutually exclusive events $E_1, E_2, \ldots, E_n$ such that

$$\bigcup_{i=1}^{n} E_i = S \tag{2.9}$$

then the E_i are said to form a *partition* of S. If $\{E_1, E_2, \ldots, E_n\}$ is a partition of the sample space S and E is any event, (which may be a subset of one of the E_i, or may include elements of several of the E_i, see Figure 2.6), then

$$P(E) = \sum_{i=1}^{n} P(E_i)P(E|E_i) \tag{2.10}$$

Using Equations (2.9) and (2.10) we may derive Bayes' law:

$$P(E_i|A) = \frac{P(A|E_i)P(E_i)}{\sum_{j=1}^{n} P(A|E_j)P(E_j)} \tag{2.11}$$

The use and power of this law may not be immediately obvious, but it underlies a large discipline of statistical literature known as Bayesian inference, which is beyond the scope of this book. This section concludes with a few examples that illustrate how Bayes' law can be used in some practical situations.

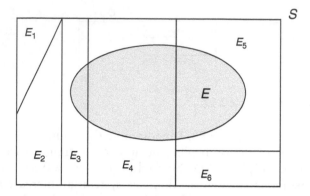

Figure 2.6 Sample space containing six mutually exclusive events $E_1, E_2, \ldots, E_6$ and an event E that includes some elements of events E_3, E_4 and E_5.

Example 2.12. Concrete units for an embankment protection scheme are manufactured by four suppliers. A supplier will be selected at random to provide the units. It is known from past experience that not all units are perfect. 5%, 40%, 10% and 10% of the units provided by Suppliers 1, 2, 3 and 4, respectively, are deficient. Furthermore, Supplier 1 can provide 2000 units, Supplier 2 can provide 500 units and Suppliers 3 and 4 can provide 1000 units each. (i) If a supplier is chosen at random, and a unit selected from their stock at random, what is the probability that the unit is defective? (ii) The selected unit is found to be defective, what is the probability that it came from Supplier 2?

Solution. (i) The sample space, S, of this experiment consists of 4000 good (g) units and 500 bad (b) units arranged as follows: Supplier 1, 1900g, 100b; Supplier 2, 300g, 200b; Supplier 3, 900g, 100b; Supplier 4, 900g, 100b. Let B_i denote the event consisting of all units from the ith supplier and D the event consisting of all defective units. Evidently, $P(B_i) = 1/4$ for all i, as the suppliers are selected at random. The conditional probabilities of selecting a defective unit, given a particular supplier, are determined directly from the ratios of defective units to the total number of units in stock at each supplier. Thus:

$$P(D|B_1) = 100/2000 = 0.05; P(D|B_2) = 200/500 = 0.4; P(D|B_3)$$
$$= P(D|B_4) = 100/1000 = 0.1.$$

Since the B_i form a partition of S we have

$$P(D) = 0.05 \times 1/4 + 0.4 \times 1/4 + 0.1 \times 1/4 + 0.1 \times 1/4 = 0.1625.$$

(ii) We require the conditional probability $P(B_2|D)$. As $P(D) = 0.1625$, $P(D|B_2) = 0.4$ and $P(B_2) = 0.25$ we find from Equations (2.10) and (2.11), $P(B_2|D) = 0.4 \times 0.25/0.1625 = 0.615$.

This result may seem rather strange at first. The chance of selecting a defective unit from Supplier 2 is 1/4. But if we have selected a defective unit then the chance that it came from Supplier 2 more than doubles to 0.615. The frequency interpretation is as follows: if the experiment is repeated n times then Supplier 2 will be selected $0.25n$ times. If n_D experiments result in a defective unit being selected, then the number of times the unit is taken from Supplier 2 equals $0.615n_D$.

Example 2.13. A court is investigating the occurrence of a damaging flood. It is alleged that this was due to inappropriate design of a flood defence structure. Let F be the event that inappropriate design occurred. Two independent expert witnesses, Bill and Ben, are advising the court. The court knows the reliability of both: Bill tells the truth with probability p_1 and Ben with probability p_2, and their statements may be taken as being independent. Let B_1 and B_2, respectively, be the events that Bill and Ben maintain that F occurred, and let $q = P(F)$. What is the probability that F occurred given that both Bill and Ben state that F occurred?

Solution. In symbols, we must evaluate $P(F|B_1 \cap B_2)$. From Equation (2.6) this is equivalent to $P(F \cap B_1 \cap B_2)/P(B_1 \cap B_2)$. However, $P(F \cap B_1 \cap B_2) = P(B_1 \cap B_2|F) \times P(F)$ and from Equation (2.10)

$$P(B_1 \cap B_2) = P(B_1 \cap B_2|F)P(F) + P(B_1 \cap B_2|F')P(F')$$

We have from the independence of the expert witnesses that B_1 and B_2 are conditionally independent given either F or F'. Thus,

$$P(B_1 \cap B_2|F) = P(B_1|F)P(B_2|F) = p_1p_2$$
$$P(B_1 \cap B_2|F') = P(B_1|F')P(B_2|F') = (1 - p_1)(1 - p_2)$$

so that, from Equation (2.11),

$$P(F|B_1 \cap B_2) = \frac{p_1p_2q}{p_1p_2q + (1 - p_1)(1 - p_2)(1 - q)}$$

With values of $p_1 = 0.95$, $p_2 = 0.8$ and $q = 0.02$ we find the probability that F occurred given that both Bill and Ben state that F occurred as 0.61, which is perhaps not as large as one might have thought at the outset.

2.3   Random variables

A *random variable*, X, is a variable whose value is determined by the result of an experiment involving unpredictable or chance causes. Random variables fall into two categories: discrete and continuous.

2.3.1   *Discrete random variables*

A variable is discrete if the set of all values it may take is countable, that is, the number of values is finite or can be put into one-to-one correspondence with the positive integers. An example of a discrete random variable is the sequence of values obtained by repeatedly throwing a die. Another example would be the readings of a tide board recorded to the nearest 0.1 m with readings limited between a and b, where a is the lowest recordable level (at or near the point where the tide board enters the seabed) and b is the (finite) upper limit on the tide board, probably Highest Astronomical Tide plus a couple of metres to account for any surge component. An example of such a tide board is shown in Figure 2.7.

Figure 2.7 Tide boards either side of lock gates at Silloth, UK, used for record-ing water levels. This board, plus another on the other side of the lock gate, were the first metric boards to be installed in the UK. (Photograph courtesy of Dr Dominic Hames.)

Table 2.1 A frequency table of water-level measurements

Water level (m)	Frequency	Probability (to 2 decimal places in brackets)	Cumulative distribution
0.5	2	0.011494 (0.01)	0.011494
0.6	4	0.022989 (0.02)	0.034483
0.7	3	0.017241 (0.02)	0.051724
0.8	7	0.040230 (0.04)	0.091954
0.9	6	0.034483 (0.03)	0.126437
1.0	5	0.028736 (0.03)	0.155173
1.1	8	0.045977 (0.05)	0.201150
1.2	15	0.086207 (0.09)	0.287357
1.3	14	0.080460 (0.08)	0.367817
1.4	18	0.103448 (0.10)	0.471265
1.5	17	0.097701 (0.10)	0.568966
1.6	16	0.091954 (0.09)	0.660920
1.7	14	0.080460 (0.08)	0.741380
1.8	11	0.063218 (0.06)	0.804598
1.9	8	0.045977 (0.05)	0.850575
2.0	11	0.063218 (0.06)	0.913793
2.1	7	0.040230 (0.04)	0.954023
2.2	6	0.034483 (0.03)	0.988506
2.3	2	0.011494 (0.01)	1.000000
	Total 174	Total 1.000000 (0.99)	

Although one might consider that water levels vary continuously, the practical limits on the range of measurable water levels mean that the values recorded will be in a finite range, and the resolution of the recordings means the set of possible recorded values will be discrete. Given a sequence of values of a discrete random variable, X, one may construct a frequency table summarising the number of times each value occurred during the course of the sequence. Table 2.1 is just such a tabulation of 174 water-level measurements constrained to lie between +0.5 m and +2.3 m, in intervals of 0.1 m.

The first column shows the water levels, the second the number of times that water level occurred in the sequence, and the third column shows the probability determined by dividing the number of occurrences by the total number of observations. The sample space is the set of values in the first column, and the events of these values being recorded at any time are mutually exclusive. Thus, from Equation (2.4), we would expect the sum of the probabilities to be 1. Column 3 shows one of the practical problems that can be encountered when deriving probabilities from a set of measurements. The probabilities are shown to six decimal places as calculated in a spreadsheet. Also shown (in brackets) are the probabilities obtained by rounding the more accurate probabilities to two decimal places. In this case, the probabilities do not sum to 1 exactly. Equation (2.4) is

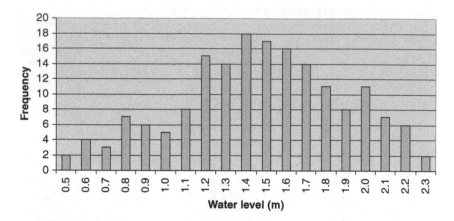

Figure 2.8 Probability function (or frequency chart) for the water levels given in Table 2.1.

obeyed approximately, and will approach fulfilment as the accuracy of the calculation of the probabilities increases. Discrepancies can arise because calculations (whether done by hand, by desk calculator or on a computer) are performed to a finite accuracy.

Table 2.1 defines the *probability function* of the random variable, X. The probability function of a discrete random variable, X, is defined as the set of pairs $\{x_i, f(x_i)\}$, where x_i is a real number, $i = 1, 2, \ldots, n$, $f(x_i)$ is the probability that $X = x_i$, and $\sum_{i=1}^{n} f(x_i) = 1$. Figure 2.8 shows the graph of the probability function of the random variable defined by Table 2.1.

The probability function is sometimes referred to as the *probability mass function* to distinguish it from the probability density function used for continuous random variables. In Table 2.1, if we now consider not the frequency that a particular value occurs, but the frequency that X takes on a value less than or equal to a particular value, we derive a cumulative probability table that defines the cumulative distribution function of X. The fourth column in Table 2.1 contains the cumulative probabilities. We write $F(x_i) = P(X \leq x_i)$, and $f(x_i) = P(X = x_i)$. Thus, $F(x_1) = f(x_1)$, $F(x_2) = f(x_1) + f(x_2)$, ..., and $F(x_n) = P(X \leq x_n) = \sum_{i=1}^{n} f(x_i) = 1$. Figure 2.9 shows the graph of the cumulative frequency function. This can be converted into the cumulative distribution function by dividing all the frequencies by the total number of observations (174).

Note that this is an increasing function for obvious reasons. The value of the cumulative distribution function $F(x_i)$ gives the probability of the random variable, X, being less than or equal to x_i. This turns out, perhaps surprisingly at this stage, often to be a more helpful quantity to deal with than the probability function $f(x_i)$.

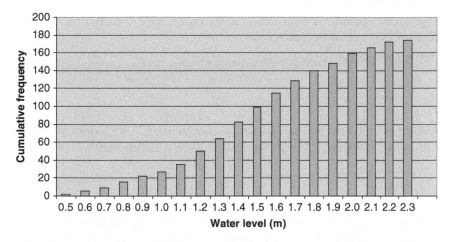

Figure 2.9 Cumulative frequency function of the water level data in Table 2.1.

Examples of discrete variables

1 Bernoulli trials: X takes values 0 or 1 with probabilities p and q, respectively. The probability function is $f(0) = p$, $f(1) = q = 1 - p$. An example is tossing a fair coin where $p = q = 1/2$.

2 Binomial distribution: If n independent Bernoulli trials $X_1, X_2, \ldots, X_n$ are performed and the total number of 1s are counted so that $Y = X_1 + X_2 + \ldots + X_n$ then the probability function of Y is

$$f(k) = {}^nC^k p^k (1-p)^{n-k}, \qquad k = 0, 1, \ldots, n \tag{2.12}$$

This is sometimes described in shorthand as 'Y is a $B(n, p)$ variable'.

3 Poisson distribution: A Poisson variable is a random variable that has the mass function

$$f(k) = \frac{\lambda^k}{k!} e^{-\lambda}, \qquad k = 0, 1, \ldots, \tag{2.13}$$

for some $\lambda > 0$. This is a limiting case of the Bernoulli function when n becomes very large and p becomes very small. It gives the probability of a number of occurrences of a random event in a given time interval, given the mean rate of occurrence, λ.

4 Geometric distribution: Suppose independent Bernoulli trials are performed at times $1, 2, \ldots$. Let Z be the time that elapses before the first success. Z may be considered the 'waiting time'. Then

$$P(Z = k) = f(k) = p(1-p)^{k-1} \tag{2.14}$$

2.3.2   *Continuous random variables*

A random variable is continuous if its cumulative distribution function $F(x) = P(X \le x)$ can be written as

$$F(x) = \int_{-\infty}^{x} f(u)du \tag{2.15}$$

where f is called the *probability density function* of X. This does not define F uniquely; however, if F is differentiable, then it is customary to write

$$f(u) = \frac{dF}{du} \tag{2.16}$$

Note that there is a correspondence between the *cumulative distribution functions* for both discrete and continuous cases. Indeed, it is nowadays customary to refer to $F(x)$ as the *distribution function*. The correspondence between the probability mass function and the probability density function is rather more complicated. The probability of a continuous random variable having a value between x and $x + \delta x$ is $F(x + \delta x) - F(x) \approx f(x)\delta x$, not $f(x)$, as in the discrete case. The numerical value $f(x)$ is not a probability. In fact, for a continuous variable $P(X = x) = 0$. This may seem nonsensical, since X needs to take some value. Broadly speaking, for a continuous variable there is an uncountable number of values it can take, and this number is so large that the probability of X taking any particular value is zero. This difference is emphasised by referring to the *probability mass function* in the discrete case and the *probability density function* for continuous variables. To convert the 'density' into a 'mass' it is necessary to multiply by a length δx. The crucial point is, that for continuous random variables, to obtain a probability it is necessary to consider a small range of values, (say x to $x + \delta x$), and multiply the probability density function value at x by the range δx to obtain the probability that X lies in the range $[x, x + \delta x]$.

In practice, the types of variables encountered in hydraulic and coastal engineering are generally considered to be continuous. However, when they are measured this can only be to a finite accuracy. For analytical manipulation it is much more convenient to work with continuous distributions. Thus, the discrete measured values are treated *as though* they were truly continuous. As an example consider Table 2.1, which contains a discrete distribution of values of an essentially continuous process. These values define a discrete distribution. [The value of the probability function is the probability of the random variable taking that value; so $f(0.5) = P(x = 0.5) = 0.011494$]. This discrete distribution can be 'transformed' into a continuous distribution by equating the probability of the discrete variable $f(x)$ to the product of a probability density function and a length. In this case the length will be the resolution of the measurements ($= 0.1\,m$). This process is often

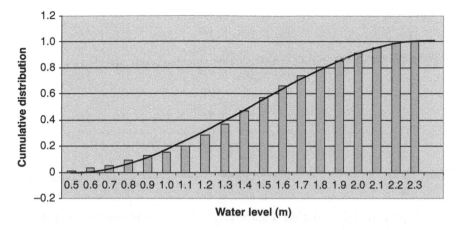

Figure 2.10 Illustration of fitting a smooth function to the discrete cumulative distribution function with the data in Table 2.1.

referred to as 'binning'. It is common practice to centre the bins about the discrete values. Thus, $f(x) = P(x - \delta x/2 \leq x < x + \delta x/2)$, where δx is the bin size and so, for example, the discrete distribution that has $P(x = 0.5)$ is interpreted as a continuous distribution with $P(0.45 \leq x < 0.55)$. The resulting probability density function has a step-like appearance. Although continuous, this form is not very convenient because the vertical steps give rise to a stepped cumulative distribution function. It is then very difficult to obtain the probability density function from this via Equation (2.16). Very often a further step is taken whereby a smoothly varying probability density function (with a convenient analytical form) will be fitted to the continuous step function. The process is illustrated in Figure 2.10.

The difference between discrete and continuous variables and between probability mass functions, probability density functions and cumulative distribution functions may have been laboured somewhat, but an incomplete understanding of the concepts and terms is the source of many errors.

Examples of continuous variables

1 Uniform or rectangular distribution (Figure 2.11): X is uniformly distributed on the range $[a, b]$ if

$$F(x) = \begin{cases} 0 & \text{if} \quad x \leq a \\ (x-a)/(b-a) & \text{if} \quad a < x \leq b \\ 1 & \text{if} \quad x > b \end{cases} \tag{2.17}$$

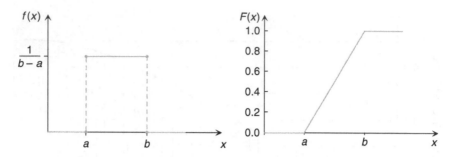

Figure 2.11 Plot of the rectangular or uniform density and distribution functions.

2 Exponential distribution (Figure 2.12): X is exponential with parameter $\lambda (> 0)$ if

$$F(x) = 1 - e^{-\lambda x}, \qquad x \geq 0 \tag{2.18}$$

This distribution is the continuous version of the waiting time or discrete geometric distribution. It often appears in practice to describe the time elapsing between unforeseeable events such as storms or earthquakes.

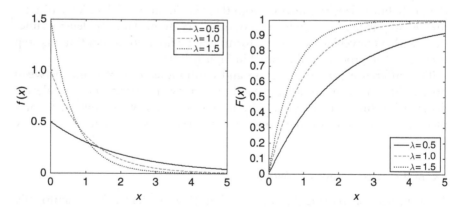

Figure 2.12 Plots of the exponential density and distribution functions for three values of the parameter λ.

3 Normal or Gaussian distribution (Figure 2.13): The normal or Gaussian distribution arises as a limiting case of other probability distributions and as the probability distribution of a sum of random processes (see

Section 3.4). The probability density function and cumulative distribution is given by

$$f(x) = \frac{1}{\sqrt{2\pi\sigma^2}} \exp\left[-\frac{1}{2}\left(\frac{x-\mu}{\sigma}\right)^2 \right] \qquad -\infty < x < \infty \qquad (2.19)$$

and

$$P(X \leq x) = F(x) = \frac{1}{\sqrt{2\pi}} \int_{-\infty}^{z} e^{-\frac{t^2}{2}} dt \qquad -\infty < x < \infty \qquad (2.20)$$

where $z = (x - \mu)/\sigma$ and t is a dummy integration variable. There is no simple expression for $F(x)$. The standard normal distribution, written $N(0, 1)$, has zero mean and unit variance. An $N(\mu, \sigma)$ variate is standardised as $z = (x - \mu)/\sigma$; and $f(z)$ and $F(z)$ are commonly denoted by $\varphi(z)$ and $\Phi(z)$, respectively. Tables of values of $\Phi(z)$ can be found in standard texts (e.g., Abramowitz & Stegun 1964). Tabulated values of $\Phi(z)$, $\varphi(z)$ and $1 - \Phi(z)$ are provided in Appendix A. The normal distribution has the following useful properties:

$$\Phi(-y) = 1 - \Phi(y)$$
$$y = \Phi^{-1}(p) = -\Phi^{-1}(1 - p) \qquad (2.21)$$
$$P(a < x \leq b) = \Phi\left(\frac{b-\mu}{\sigma}\right) - \Phi\left(\frac{a-\mu}{\sigma}\right)$$

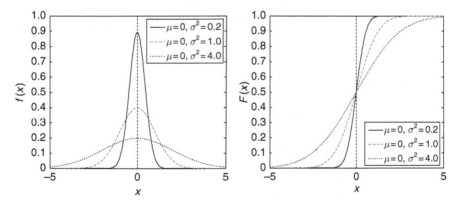

Figure 2.13 Plots of the normal density and distribution functions, for zero mean and different values of the variance.

Example 2.14. Illustrative calculations using the normal distribution tables (Appendix A). From the tables in Appendix A find:

$$\text{(a) } \int_{0.50}^{\infty} \varphi(z)dz, \text{ (b) } \int_{2.17}^{\infty} \varphi(z)dz, \text{ (c) } \int_{1.50}^{2.0} \varphi(z)dz, \text{ (d) } \int_{-1.50}^{\infty} \varphi(z)dz, \text{ (e) } \int_{-1.32}^{2.13} \varphi(z)dz$$

Solution.

a   From Table A we find $\Phi(0.5) = \Phi(0.5) = \int_{-\infty}^{0.5} \varphi(z)dz = 0.69146$. Thus

$$\int_{0.50}^{\infty} \varphi(z)dz = 1 - \Phi(0.5) = 0.3085.$$

b   We can use Table A as in part (a) or use Table D directly to find

$$\int_{2.17}^{\infty} \varphi(z)dz = 0.0150.$$

c   Note that

$$\int_{1.50}^{2.0} \varphi(z)dz = \int_{-\infty}^{2.0} \varphi(z)dz - \int_{-\infty}^{1.5} \varphi(z)dz = \Phi(2.0) - \Phi(1.5)$$

$$= 0.97724 - 0.93319 = 0.04405.$$

d   Table D does not extend to negative arguments (otherwise we could use this directly), so we note that

$$\int_{-1.50}^{\infty} \varphi(z)dz = 1 - \int_{-\infty}^{-1.50} \varphi(z)dz = 1 - \int_{1.50}^{\infty} \varphi(z)dz = 1 - 0.0668 = 0.9332.$$

e   Note that

$$\int_{-1.32}^{2.13} \varphi(z)dz = 1 - \left[\int_{-\infty}^{-1.32} \varphi(z)dz + \int_{2.13}^{\infty} \varphi(z)dz \right]$$

$$= 1 - \left[\int_{1.32}^{\infty} \varphi(z)dz + \int_{2.13}^{\infty} \varphi(z)dz \right]$$

$$= 1 - (0.0934 + 0.0166) = 0.8900, \text{ using Table D.}$$

4   Rayleigh distribution (Figure 2.14): X follows a Rayleigh distribution
with parameter b if

$$F(x) = 1 - \exp\left[-\frac{x^2}{2b^2}\right], \qquad 0 \le x \tag{2.22}$$

Under suitable conditions, the distribution of wave heights can be described
with a Rayleigh distribution.

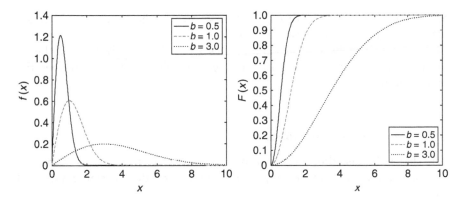

Figure 2.14 Plot of the Rayleigh density and distribution functions for three values
of the parameter b.

5   Gamma distribution (Figure 2.15): X follows a Gamma distribution
with parameters λ and r if

$$F(x) = \int\limits_0^x \frac{\lambda^r w^{r-1} e^{-\lambda w}}{\Gamma(r)} dw, \qquad 0 \le x, \lambda > 0, r > 0 \tag{2.23}$$

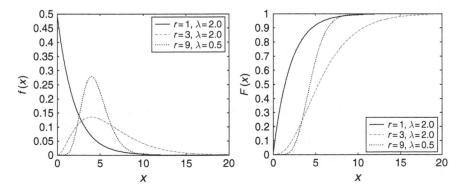

Figure 2.15 Plots of the gamma density and distribution functions for a range of
parameter values.

where w is a dummy variable, and $\Gamma(r)$ is the complete Gamma function defined as $(r-1)!$ when r is an integer. Where r is not an integer

$$\Gamma(r) = \begin{cases} \int\limits_0^\infty t^{r-1}e^{-t}dt & r > 0 \\ 0 & otherwise \end{cases}$$

When r is an integer this distribution is known as the Erlang distribution, after the Danish engineer who derived it in his study of congestion on telephone lines. The Erlang distribution describes the distribution of the rth arrival of a Poisson process with parameter λ, which is usually termed the 'scale parameter'.

2.3.3 *Moments of a distribution*

The shape of a distribution may be described quantitatively by its moments. The *mean*, or *expected value* or *first moment* of a random variable is defined as:

$$E(X) \equiv \mu = \int\limits_{-\infty}^\infty xf_X(x)dx \approx \sum_i x_i p_X(x_i) \tag{2.24}$$

The integral expression is for continuous random variables and the summation expression is for discrete ones. The name 'first moment' arises because it is the first moment of 'area' of the probability density function about the origin.

If n is a positive integer, then the nth *central moment*, M_1, of X is

$$M_n = E[(X - M_1)^n] \tag{2.25}$$

where $M_1 = \mu$. The central moments that are most widely used are the first and second, which are the mean and the *variance*, respectively. The positive square root of the variance is called the *standard deviation*, and is often written as σ. Moments may also be defined without respect to the mean. Thus, the *ordinary moments*, m_i, of X are calculated as:

$$m_n = E(X^n) \tag{2.26}$$

The central moments may be written in terms of the ordinary moments. As an example,

$$
\begin{aligned}
M_2 &= E[(X - m_1)^2] \\
&= E(X^2) - E(2m_1 X) + E(m_1^2) \\
&= m_2 - 2m_1 E(X) + E(m_1^2) \\
&= m_2 - m_1^2
\end{aligned}
\tag{2.27}
$$

In passing, it should be noted that variances can *never* be negative.

Higher moments are sometimes calculated as a test of normality. The two most commonly used are the third and fourth central moments. A measure of the asymmetry about the mean of a distribution is given by the *skewness*. This is defined as:

$$
Skewness = \frac{E[(X - \mu)^3]}{\sigma^3}
\tag{2.28}
$$

A normal distribution is symmetric about the mean and has zero skewness. A positive value of skewness indicates that the density function has a longer 'tail' in the positive direction. A measure of the 'peakedness' of a distribution is given by the *kurtosis*, defined as:

$$
Kurtosis = \frac{E[(X - \mu)^4]}{\sigma^4}
\tag{2.29}
$$

A standard normal distribution has kurtosis equal to 3.0. A kurtosis less than 3 indicates a density function that is more sharply peaked than a normal density function, while values greater than 3 are symptomatic of a broader peak than that of a normal density function.

Example 2.15. Calculate the mean and variance of the uniform variate X that takes values between a and b inclusive.

Solution. The density function for X is $1/(b-a)$ for $a \le X \le b$. Thus

$$
E(X) = \int_{-\infty}^{\infty} \frac{x}{b-a} dx = \int_{a}^{b} \frac{x}{b-a} dx = \frac{b+a}{2}
$$

$$
Var(X) = \int_{a}^{b} \frac{x^2}{b-a} dx - [E(X)]^2 = \frac{(b-a)^2}{12}
\tag{2.30}
$$

More generally, the mean of the random variable $Y = g(X)$, where g is an arbitrary function of x, is given by:

$$E[g(X)] = \int_{-\infty}^{\infty} g(x)f_X(x)dx \qquad (2.31)$$

2.3.4   *Percentiles, quartiles and quantiles*

A concept that can be very useful is that of a percentile. The mth quantile of a random variable X is the variate that corresponds to the cumulative distribution function value m. The mth quantile is also known as the $100m$th percentile. Quartiles are a special case of percentiles: the first quartile, Q_1, is the 25th percentile or the value of the variable for which $F_X(x) = 0.25$; the second quartile, Q_2, is the 50th percentile or median; and the third quartile, Q_3, is the 75th percentile. Quartiles and percentiles are of particular use when applied to symmetric distributions such as the normal distribution. For the $N(0, 1)$ distribution, the area under the curve from $z = 1$ to $z = \infty$ is 0.1587. [By symmetry this area is the same as the area under the curve from $z = -\infty$ to $z = -1$. This area is equal to $\Phi(-1)$. From Equation (2.21), $\Phi(-1) = 1 - \Phi(1)$, giving the result as $\Phi(1) = 0.8413$ from Table A in Appendix A.] Therefore, the area under the curve between $z = -1$ to $z = +1$ is $1 - 2 \times 0.1587 = 0.6826$, that is, about 68% of normal variates deviate from their mean by less than one standard deviation. Repeating the calculations for $z = \pm2$ and ±3 we find that 95.4% and 99.7% of the area under the curve is included between the respective limits, as shown in Figure 2.16. The quartiles divide the area under the curve into quarters. The values of z corresponding to the 1st and 3rd quartiles can be found by reverse interpolation from Table A2 to be ±0.6745, respectively. Thus, 50% of the area under the curve lies in the interval $-0.6745 \leq z \leq 0.6745$.

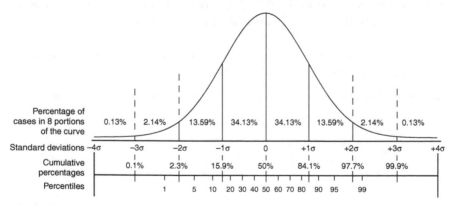

Figure 2.16 Illustration of the standard normal quantiles. The 'x-axis' denotes values of z.

2.3.5 *Transformation of continuous distributions*

If $g(\ )$ is a monotonic function and has an inverse, so that if $Y = g(X)$ and $X = g^{-1}(Y)$, and X and Y are random variables, then the probability density function of Y is given by

$$f_Y(y) = \frac{f_X(x)}{\left|\frac{dg}{dx}\right|} \tag{2.32}$$

Note that subscripts on the functions have been used to distinguish which probability density function is meant. Thus, $f_Y(y)$ is the value of the probability density function for Y taken when the value y is used. If g is not monotonic Equation (2.32) is no longer applicable. The probability density function of the transformed variable may still be found, but must be determined on a case-by-case basis (see Section 3.1).

Example 2.16.

1   $Y = aX + b$. The equation $y = g(x) = ax + b$ has a single solution $x = (y - b)/a$ for each y. Hence

$$f_Y(y) = f_X\left(\frac{y-b}{a}\right)\left|\frac{1}{a}\right| \tag{2.33}$$

2   In case 1, if X is uniform in the interval $x_1 < x < x_2$, then Y is uniform in the interval $ax_1 + b < y < ax_2 + b$.

3   If we take $y = g(x) = F_X(x)$, $0 \leq y \leq 1$. Assuming F_X is differentiable, $dg/dx = dF_X(x)/dx = f_X(x)$. Substituting into Equation (2.32) gives

$$f_Y(y) = f_X(x)\left|\frac{1}{f_X(x)}\right| \qquad 0 \leq y \leq 1 \tag{2.34}$$

This is a rectangular distribution, irrespective of the form of $F_X(x)$.

2.3.6 *Numerical generation of random variables*

It is reasonable to ask: 'How do you go about actually creating a sequence of random numbers or random variables?'. We start by considering how to generate random numbers that lie uniformly in the interval [0, 1], and then discuss how these can be used to generate random variables with other distributions.

One of the more usual methods is to use a 'pseudo' random number generator, or PRNG. These routines are available on most computers either as part of specialist software packages or as part of more everyday software. It is important to recognise that such routines do not produce a truly random sequence of numbers. As a digital computer works to a fixed precision, there is in fact a finite number of different values that it can represent. A good PRNG will sample this finite set of numbers randomly. However, repetition will occur eventually. Indeed, most standard PRNGs cycle through a pre-determined sequence of numbers. The initial starting point on a sequence is determined by the 'seed' value given to the routine when it is first called. A well-designed PRNG can have a very long sequence before repetition, of the order of tens of millions of elements. The algorithm at the core of many PRNGs is the linear congruential generator. A brief outline of the method is given here. For a further discussion of the advantages and disadvantages of this technique the reader is referred to Press et al. (2007) and references therein.

A *linear congruential generator* is determined by three positive integers a, b and m, where a is the *multiplier*, b the *increment* and m the *modulus*. It produces its pseudo-random sequence of integers $\{n_k\}(0 \le n_k < m)$ as follows:

- specify n_0 (the *seed*); and
- generate the rest by

$$n_{k+1} = an_k + b \;(\text{mod } m) \tag{2.35}$$

- then set $x_k = n_k/m$ so that $0 \le x < 1$.

In practice, m is usually something like

$$2^{32} = 4,294,967,296$$

and a is chosen carefully to give good results. (The choice of a is critical; the choice of b is less important.)

In order to have a good generator the following conditions should be satisfied (Knuth 1981):

- The modulus m should be large (not a major obstacle for computer arithmetic).
- Taking m to be a power of 2, pick a such that $a \bmod 8 = 5$. This will have the effect of maximising the period of the necessarily cyclic sequence produced. The choice of value for a is the most delicate part of the whole design.
- The increment b should be co-prime to m.

An acceptable choice of values would be: $m = 2^{32}$, $a = 1,664,525$ and $b = 1$. In the above, if $b = 0$ in Equation (2.33), the algorithm is called *a multiplicative linear congruential generator*.

Example 2.17. As a simple example, let $m = 2^3$, $a = 5$ and $b = 1$. Taking $n_0 = 1$ we find

$$n_1 = 5 \times 1 + 1 \,(\mathrm{mod}\,8)$$
$$= 6 \,(\mathrm{mod}\,8)$$
$$= 6$$

The seed for the next random number is n_1 and the random number is $n_1/8 = 0.750$. The first nine terms of this sequence are given in Table 2.2.

With such a small value of m the cycle of random numbers will be very small and the coverage of all the random numbers between 0 and 1 will be very coarse (at intervals of 1/8). Indeed, the cycle repeats after only 8 steps. If we chose a different starting seed, say $n_0 = 2$, then we just start the process at a different point in the cycle. A good algorithm will make full use of the bit-length of a particular computer, thereby maximising the length of the cycle.

Most applications involve variables that do not obey a uniform distribution. In order to generate sequences of values that obey non-uniform distributions we can use some of the earlier results. The most general technique is the 'inverse transformation' method. The method is as follows. Given a random variable X_i, its cumulative distribution function, $F_{Xi}(x_i)$, lies in the range [0, 1]. Given a sequence of uniformly distributed random numbers between 0 and 1, r_i, we can generate a corresponding sequence of transformed variates by setting $X_i = F_{Xi}^{-1}(r_i)$. Figure 2.17 illustrates how the 'inverse transformation' method works. For cases where F_{Xi}^{-1} can be expressed in closed analytical form the inverse transform can be a very efficient means of generating nonuniformly distributed variables. For cases where no closed form exists, the method still works, but the

Table 2.2 Illustration of the linear congruential random number generation algorithm

Term (k)	n_k	x_k
0	1	–
1	6	0.750
2	7	0.875
3	4	0.500
4	5	0.625
5	2	0.250
6	3	0.375
7	0	0.000
8	1	0.125
9	6	0.750

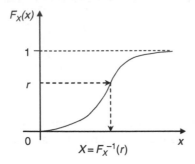

Figure 2.17 Illustration of the 'inverse transformation' technique for generating random numbers with a specified distribution. Uniformly distributed random numbers r, lying between 0 and 1, are generated and then converted to variates with distribution F_X through inverse transformation.

inverse function has to be evaluated graphically, by numerical integration, by tables and interpolation, or by fitting an approximating function to tabulated values.

Example 2.18. Find the inverse transform function for the exponential distribution.

Solution. We set the random number $r = F(x)$ and solve for x:

$$F(x) = 1 - e^{-\lambda x}$$
$$\therefore \quad r = 1 - e^{-\lambda x}$$
$$\therefore \quad 1 - r = e^{-\lambda x} \tag{2.36}$$
$$\therefore \quad \ln(1 - r) = -\lambda x$$
$$\therefore \quad x = -\frac{\ln(1 - r)}{\lambda}$$

Example 2.19. Find the inverse transform function for the Rayleigh distribution.

Solution. As in the previous example, we solve for x:

$$F(x) = 1 - e^{-x^2/(2b^2)}$$
$$\therefore \quad r = 1 - e^{-x^2/(2b^2)}$$
$$\therefore \quad 1 - r = e^{-x^2/(2b^2)} \tag{2.37}$$
$$\therefore \quad \ln(1 - r) = -x^2/(2b^2)$$
$$\therefore \quad x = \sqrt{-2b^2 \ln(1 - r)}$$

For $0 \leq r < 1$, the expression under the root sign is positive. The required value of x is obtained by taking the positive square root as the Rayleigh distribution is defined only for $x \geq 0$.

2.4   Reliability and hazard functions

We conclude this chapter with a short section on some *functions* of probability density and cumulative distribution functions. These appear sufficiently frequently in some reliability problems that they have been defined in their own right. The first is the *survival function*, which, for a random variable, X, is defined as:

$$G(x) = P(X > x) = 1 - F(x) \tag{2.38}$$

This terminology has arisen because if X is considered to be the strength of an object subjected to loading, then the survival function is simply the probability that the strength of the object is greater than the imposed load, i.e., the probability that the object survives the loading. The survival function is also termed the *reliability function*, for obvious reasons. It should be noted that a slightly modified definition of the reliability function is used in reliability theory and in this book in Chapters 5, 6 and 7. The modified version is simply the difference between the strength and the load.

The *hazard function*, $h(x)$, is the probability of failure given that no failures have occurred so far:

$$h(x) = \frac{f(x)}{1 - F(x)} \tag{2.39}$$

The hazard function is also termed the *failure rate* or *hazard rate*. It is most often encountered in problems concerning time variation (x replaced by t in Equation 2.39), with the interpretation that the hazard function is the probability that failure occurs for $t \leq T \leq t + \delta t$ given that there have been no failures for $t \leq T$.

The cumulative or integrated hazard function is the integral of the hazard function:

$$H(x) = \int_{-\infty}^{x} h(u)du = -\ln[1 - F(x)] = -\ln[G(x)] \tag{2.40}$$

Equation (2.40) provides a neat relation between the four different functions.

Example 2.20. Find the reliability, hazard and integrated hazard function for the exponential distribution.

Solution. For the exponential distribution we have $F(x) = 1 - e^{-\lambda x}$ and $f(x) = \lambda e^{-\lambda x}$. Thus $G(x) = 1 - F(x) = e^{-\lambda x}$; $h(x) = \lambda$; and $H(x) = -\log(e^{-\lambda x}) = \lambda x$.

Returning to the case where the argument of the hazard function is time, note that the exponential distribution has a constant hazard function. Proneness to failure at any time is constant, and is not time dependent. This is suitable for elements that do not age, for example, semi-conductor components.

If $h(t)$ is an increasing function of t, the T is said to have an increasing failure rate. This is appropriate for items that age or wear. If $h(t)$ is a decreasing function of time, then T has a decreasing failure rate. This could happen when a manufacturing process produces low-quality units so that many will fail early. If $h(t)$ has a U or 'bathtub' shape, this corresponds to high early failure rate, followed by a period of stability, followed by a wear-out period. Such a shape has been proposed as a possible model for the life distribution of humans.

A flexible form of hazard function is:

$$h(t) = \alpha \mu^{-\alpha} t^{\alpha - 1} \tag{2.41}$$

where α, $\mu > 0$ are shape and location parameters. $\alpha < 1$ gives a decreasing hazard, $\alpha = 1$ gives a constant hazard and $\alpha = 2$ gives a linearly increasing hazard. We can work out the probability density and cumulative distribution functions for this hazard function. From Equation (2.40) we have:

$$H(t) = \int_{-\infty}^{t} h(u) du = -\ln[1 - F(t)]$$

$$\therefore\ 1 - F(t) = e^{-\int h(t)dt} = e^{-\alpha \mu^{-\alpha} \int t^{\alpha - 1} dt} = e^{-\left(\frac{t}{\mu}\right)^{\alpha}}$$

$$\therefore\ F(t) = 1 - e^{-\frac{t^{\alpha}}{\mu}} \tag{2.42}$$

$$\therefore\ f(t) = \frac{d}{dt}[F(t)] = \alpha \mu^{\alpha} t^{\alpha - 1} e^{-\left(\frac{t}{\mu}\right)^{\alpha}}$$

This particular distribution is well known and is called the two-parameter *Weibull distribution*. Note that for $\alpha = 1$ it reduces to the exponential distribution with mean μ, and for $\alpha = 2$ it reduces to the Rayleigh distribution.

Example 2.21. Concrete pipe sections for a drain are known to have a Weibull failure density, with $a = 2$ and $b = 300,000$. What is the probability that such a component survives longer than 5 years?

Solution. Firstly, 5 years $= 24 \times 365.25 \times 5$ hours $= 43,830$ hours. The required probability is given by the value of the reliability function, $G(t = 43,830, a = 2, b = 300,000)$. Thus,

$$G(t = 43830, a = 2, b = 300000 = e^{-\left(\frac{t}{b}\right)^a} = e^{-\left(\frac{43830}{300000}\right)^2} \approx 0.9789$$

Associated with the concept of reliability is the mean time to failure, (MTTF), illustrated in Figure 2.18. The reliability is the probability of no failures over a stated time interval. Suppose a single item is characterised by the distribution function $F(t) = P(\tau \leq t)$ of its failure-free operating time, τ. Its reliability function, $G(t)$, is given by

$$G(t) = P[\text{no failure in}(0, t)] = 1 - F(t) \tag{2.43}$$

The MTTF can be computed from Equation (2.24), as

$$MTTF = \int_0^\infty t f(t) dt$$

$$= [tF(t)]_0^\infty - \int_0^\infty F(t) dt \tag{2.44}$$

$$= [t(1 - G(t))]_0^\infty - \int_0^\infty (1 - G(t)) dt$$

$$= [t(1 - G(t)) - t]_0^\infty + \int_0^\infty G(t) dt$$

$$= [tG(t)]_0^\infty + \int_0^\infty G(t) dt$$

$$= \int_0^\infty G(t) dt$$

where we have used integration by parts, Equation (2.38), and the assumption that $G(t)$ tends to 0 faster than t tends to infinity, so that $\underset{t \to \infty}{Lim} (tG(t)) = 0$

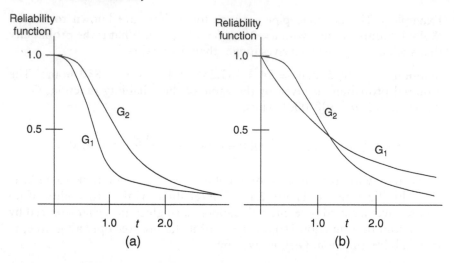

Figure 2.18 Illustrations of reliability functions G_1 and G_2, where: (a) G_2 has a larger MTTF than G_1, (b) both G_1 and G_2 have the same MTTF.

Further reading

Benjamin, J. R. and Cornell, C. A., 1970. *Probability, Statistics and Decision Theory for Civil Engineers*, McGraw-Hill, New York.

Evans, M., Hastings, N. and Peacock, B., 1993. *Statistical Distributions*, 2nd edition, Wiley, New York, p. 170.

Feller, W., 1968. *An Introduction to Probability Theory and Its Applications*, Vol. 1, 3rd edition, Wiley, New York, p. 509.

Grimmett, G. R. and Stirzaker, D. R., 1992. *Probability and Random Processes*, 2nd edition, Oxford Science Publications, Oxford, p. 541.

Papoulis, A., 1984. *Probability, Random Variables and Stochastic Processes*, 2nd edition, McGraw-Hill International Editions, Singapore, p. 576.

3 Elements of probability and stochastic processes

3.1 Manipulating distributions

In this chapter further definitions and results are presented. These are relevant to aspects of reliability theory but also introduce some ideas that are not found in most textbooks on reliability theory. The style adopted is necessarily brief in order to cover the material. Illustrative explanations of results are given but formal mathematical proofs are excluded as they may be found in other sources (see further reading at the end of this chapter).

3.1.1 Functions of one variable

A simple case of working out the probability distribution of a trans-formed random variable was covered in Section 2.3.5. The more general case, where the transforming function is not monotonic, is covered in this section. First, the result is quoted (with an illustrative explanation), and then several examples are provided to amplify the case-by-case approach required.

Finding $f_Y(y)$: It is required to determine the density function of $Y = g(X)$ in terms of the density function of X. To find $f_Y(y)$ for a specific y, solve the equation $y = g(x)$. Denoting its real roots by x_n,

$$y = g(x_1) = g(x_2) = \ldots = g(x_n) = \ldots$$

and

$$f_Y(y) = \frac{f_X(x_1)}{g'(x_1)} + \ldots + \frac{f_X(x_n)}{g'(x_n)} + \ldots \tag{3.1}$$

where $g'(x)$ is the derivative of $g(x)$.

Explanation: For sake of argument, let us assume that the equation $y = g(x)$ has two roots as shown in 3.1.

Now, $f_Y(y)dy = P\{y < Y \le y + dy\}$. Thus, we need to find the set of values x such that $y < g(x) \le y + dy$ and the probability that X is in this set.

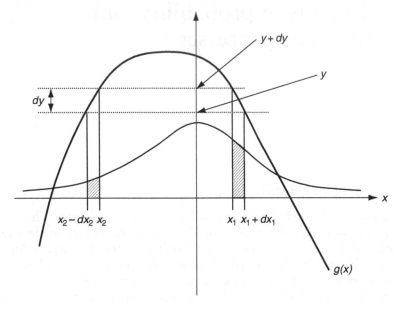

Figure 3.1 Illustration of the transformation of the density function of a function of one variable.

From Figure 3.1 it is evident that this set consists of **two** intervals: $x_1 < x < x_1 + dx_1$ and $x_2 - dx_2 < x < x_2$. Thus,

$$P\{y < Y < y + dy\} = P\{x_1 < X < x_1 + dx_1\} + P\{x_2 - dx_2 < X < x_2\}$$

The right-hand side of this expression is the sum of the shaded areas in Figure 3.1. Since

$$P\{x_1 < x < x_1 + dx_1\} = f_X(x_1)dx_1 \text{ and } dx_1 = dy/|g'(x_1)|, \text{ and}$$
$$P\{x_2 - dx_2 < x < x_2\} = f_X(x_2)dx_2 \text{ and } dx_2 = dy/|g'(x_2)|, \text{ we have that}$$

$$f_Y(y) = \frac{f_X(x_1)}{|g'(x_1)|} + \frac{f_X(x_2)}{|g'(x_2)|} \tag{3.2}$$

where dashes denote the derivative with respect to x. Taking the absolute value of the derivative of g ensures that $f_Y(y)$ integrates to unity.

Example 3.1. If X is a random variable with density function $f_X(x)$, find the density function of $Y = aX^2$ for $a > 0$.

Solution. $g'(x) = 2ax$. If $y < 0$, then the equation $y = ax^2$ has no real solution, so $f_Y(y) = 0$. If $y > 0$ then it has two solutions: $x_1 = \sqrt{(y/a)}$ and $x_2 = -\sqrt{(y/a)}$, and Equation (3.2) gives:

$$f_Y(y) = \frac{1}{2a\sqrt{y/a}}\left[f_X\left(\sqrt{\frac{y}{a}}\right) + f_X\left(-\sqrt{\frac{y}{a}}\right)\right] \tag{3.3}$$

Special case: If $a = 1$, and X is $N(0, 1)$, then Equation (3.3) gives

$$f_Y(y) = \frac{1}{\sqrt{2\pi y}}e^{-\frac{y}{2}} \tag{3.4}$$

for $y \geq 0$. This is the density function of the *chi-squared distribution* with one degree of freedom.

Example 3.2. If X is a random variable with density function $f_X(x)$, find the density function of $Y = \tan(X)$.

Solution. $g'(x) = 1/\cos^2(x) = 1 + y^2$. The equation $y = \tan(x)$ has infinitely many roots, $x_n = \tan^{-1}(y)$ with $n = \ldots, -1, 0, 1, \ldots$

$$f_Y(y) = \frac{1}{1 + y^2} \sum_{n=-\infty}^{\infty} f_X(x_n) \tag{3.5}$$

Special case: If X is uniform in $(-\pi/2, \pi/2)$ then all terms in the summation in Equation (3.5) are zero except the first, which is equal to $1/\pi$. Thus, the density function of Y is

$$f_Y(y) = \frac{1/\pi}{1 + y^2} \tag{3.6}$$

which is the density function of the *Cauchy distribution*, shown in Figure 3.2.

The Cauchy distribution is unimodal and symmetric, but has much heavier tails than the normal distribution. It is also notable for the fact that it has no moments; i.e., the integral in Equation (2.23), and similar for higher moments, is infinite.

Example 3.3. *Truncated distribution.* Let X be a random variable with density $f_X(x)$ and distribution function $F_X(x)$. If the values that X may take are limited, how does this affect its density and distribution functions?

Solution. In some instances the distribution of a random variable may be limited to an upper or lower bound. Taking the first case, let

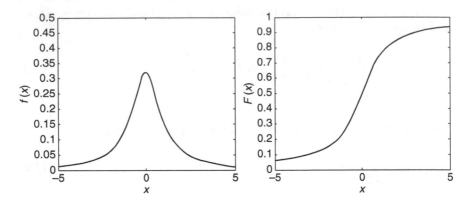

Figure 3.2 The Cauchy density and distribution function.

x_{max} be the upper limit, then the resulting distribution of the variate X subject to truncation, denoted by X', is given by the conditional density:

$$f_{X'}(x) = f_{X; X \leq x_{max}} = \begin{cases} a f_X(x) & x \leq x_{max} \\ 0 & \text{otherwise} \end{cases} \qquad (3.7)$$

where the constant a is required to normalise the density so that its integral over all values of x remains equal to 1. The constant a is given by:

$$a = \frac{1}{F_X(x_{max})} \qquad (3.8)$$

In the case where there are upper and lower bounds, x_{max} and x_{min} say, the constant a is given by:

$$a = \frac{1}{F_X(x_{max}) - F_X(x_{min})} \qquad (3.9)$$

In the specific case where X is $N(\mu, \sigma)$ the constant a takes the form

$$a = \frac{1}{\Phi\left(\dfrac{x_{max} - \mu}{\sigma}\right) - \Phi\left(\dfrac{x_{min} - \mu}{\sigma}\right)} \qquad (3.10)$$

This is illustrated in Figure 3.3 for a $N(70, 10)$ distribution curtailed below at $x = 65$ and above at $x = 85$.

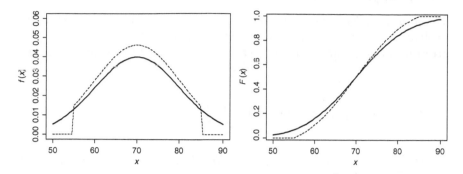

Figure 3.3 Normal, N(70, 10), and truncated density and distribution functions.

Example 3.4. If X is a random variable with density function $f_X(x)$, find the density function of $Y = aX + b$ where a and b are constants. This corresponds to a simple translation and scaling of the variable X. This function is also used for transforming variables into standard normal form.

Solution. We have $y = g(x) = ax + b$, so $x = (y - b)/a$ is the unique solution (as long as $a \neq 0$). Also, $g'(x) = a$. Thus, from Equation (3.2) we have

$$f_Y(y) = \frac{f_X\left(\dfrac{y-b}{a}\right)}{|a|} \tag{3.11}$$

Example 3.5. If X is a random variable with density function $f_X(x)$, find the density function of $Y = \sqrt{(aX)}$ where a is a constant.

Solution. We have $y = g(x) = \sqrt{(ax)}$, so $x = y^2/a$. Also, $g'(x) = \sqrt{(a/4x)}$. Thus, from Equation (3.2) we have

$$f_Y(y) = \frac{f_X\left(\dfrac{y^2}{a}\right)}{\sqrt{a/4x}} = 2\frac{y}{a}f_X\left(\frac{y^2}{a}\right) \tag{3.12}$$

Example 3.6. If X is a random variable with density function $f_X(x)$, find the density function of $Y = \sqrt{(aX + b)}$ where a and b are constants.

Solution. We can use Equation (3.2) as in the previous examples, or we can use the results from the previous examples. Let $Z = aX + b$ and $Y = \sqrt{Z}$. From Example 3.4 we have $f_Z(z) = f_X[(z - b)/a]$. From Example 3.5 we have $f_Y(y) = 2yf_Z(y^2) = 2(y/a)f_X\left[(y^2 - b)/a\right]$.

3.2   Joint probability

The *joint distribution* of two random variables X and Y, $F_{X,Y}(x,y)$, is the probability of the event $\{X \leq x, Y \leq y\} = \{(X,Y) \in D\}$, where x and y are arbitrary numbers and D is the area in the x–y plane for which $X \leq x$ and $Y \leq y$:

$$F_{X,Y}(x,y) = P[(X \leq x) \cap (Y \leq y)] = \int\limits_{-\infty}^{x} \int\limits_{-\infty}^{y} f_{X,Y}(u,v)\,du\,dv \qquad (3.13)$$

where $f_{X,Y}(x,y) \geq 0$ is the *joint probability density function*. The following conditions also hold:

$$\left. \begin{array}{l} F_{X,Y}(-\infty,-\infty) = 0 \\ F_{X,Y}(-\infty,y) = F_{X,Y}(x,-\infty) = 0 \\ F_{X,Y}(\infty,y) = F_{X,Y}(x,\infty) = 0 \\ F_{X,Y}(\infty,\infty) = 1 \end{array} \right\} \qquad (3.14)$$

The last expression is just the statement that the volume under the joint density function is unity. Similar expressions apply for discrete random variables. Where there are several random variables, the statistics of each are called *marginal*. Thus, F_X and f_X are the *marginal distribution and density*, respectively. The marginal distribution and density may be obtained from the joint distribution and density by:

$$\left. \begin{array}{l} F_X(x) = F_{X,Y}(x,\infty) \\ F_Y(y) = F_{X,Y}(\infty,y) \\ f_x(x) = \int\limits_{-\infty}^{\infty} f(x,y)\,dy \\ f_y(x) = \int\limits_{-\infty}^{\infty} f(x,y)\,dx \end{array} \right\} \qquad (3.15)$$

The conditional probability density of X and Y is given by:

$$\lim_{\delta x, \delta y \to 0} \{P(x < X \leq x + \delta x \,|\, y < Y \leq y + \delta y)\} \equiv f_{X|Y}(x|y) = \frac{f_{XY}(x,y)}{f_Y(y)} \quad (3.16)$$

The relationship between joint, conditional and marginal density functions of two variables is illustrated in Figure 3.4.

Example 3.7. A rainfall event at a recording station is defined by two variables, the duration X of the storm and its intensity Y (equal to the average rainfall rate). Suppose the two variables X and Y are jointly distributed with a *bivariate exponential* density function:

$$f_{XY}(x,y) = [(a + cy)(b + cx) - c]\,e^{-ax - by - cxy} \qquad (3.17)$$

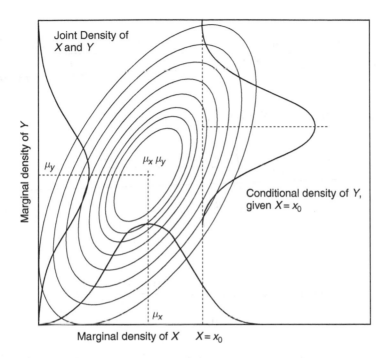

Joint Density of
X and *Y*

Marginal density of *Y*

μ_y

$\mu_x \mu_y$

Conditional density of *Y*,
given $X = x_0$

μ_x

Marginal density of *X* $X = x_0$

Figure 3.4 Relationship between joint, marginal and conditional density functions.

where a, $b > 0$, and $0 \le c \le 1$ are three parameters evaluated from the rain gauge data. Note also that x, $y \ge 0$. For now, let us take $a = 0.03\,\text{h}^{-1}$, $b = 1.5\,\text{h/mm}$ and $c = 0.05\,\text{mm}^{-1}$. It is required to find the probability that a storm lasting 6 hours will exceed an average intensity of 3 mm/h.

Solution. The probability that a storm lasting 6 hours has an intensity greater than 3 mm/h is given by $1 - F_{Y|X}(3, 6)$. To determine the conditional distribution function we need to use the density functions. Since the conditional *pdf* of the storm intensity for fixed duration is:

$$f_{Y|X}(x, y) = \frac{f_{XY}(x, y)}{f_X(x)} \tag{3.18}$$

we need to determine the marginal density $f_X(x)$. From Equation (3.15) we have

$$f_x(x) = \int_0^\infty f(x, y)\,dy = \int_0^\infty \left((a + cy)(b + cx) - c\right) e^{-ax - by - cxy}\,dy = ae^{-ax}$$

Thus

$$f_{Y|X}(x,y) = \frac{[(a+cy)(b+cx)-c]e^{-ax-by-cxy}}{ae^{-ax}} = \frac{[(a+cy)(b+cx)-c]e^{-y(b+cx)}}{a}$$

(3.19)

Hence, the conditional distribution $F_{Y|X}(x,y)$ is given by

$$F_{Y|X}(x,y) = \int_0^y f_{Y|X}(x,y)du = \int_0^y \frac{[(a+cu)(b+cx)-c]e^{-u(b+cx)}}{a}du$$

$$= 1 - \frac{a+cy}{a}e^{-(b+cx)y}$$

(3.20)

and so

$$P(Y>3|x=6) = 1 - F_{Y|X}(3,6) = 1 - 1 + \frac{0.03+0.05\times3}{0.03}e^{-(1.5+0.05\times6)3}$$

$$= 0.027$$

We now present some important definitions. Two random variables X and Y *are called independent* if

$$f_{XY}(x,y) = f_X(x)f_Y(y)$$

(3.21)

The *covariance* of two random variables X and Y, denoted Cov(X, Y), is defined as the expectation of the product between the respective deviations from their mean:

$$\text{Cov}(X,Y) = E[(X-E(X))(Y-E(Y))] = E(X,Y) - E(X)E(Y)$$

(3.22)

If X and Y are independent, then their covariance is zero. However, the reverse is not always true; just because the covariance of two variables is zero does not mean they are independent.

The (linear) *correlation coefficient* ρ is the normalised covariance between two variables:

$$\rho = \frac{\text{Cov}(X,Y)}{\sigma_X\sigma_Y}$$

(3.23)

where σ_X and σ_Y are the standard deviations of X and Y, respectively. The correlation coefficient lies between -1 and 1.

The bivariate exponential density was introduced in Example 3.7. Another well-known bivariate density is the *joint normal density*. This is defined as:

$$f_{XY}(x,y) = \frac{1}{2\pi\sigma_X\sigma_Y\sqrt{1-\rho^2}} e^{-\left(\frac{1}{2(1-\rho^2)}\left[\frac{(x-\mu_X)^2}{\sigma_X^2} - 2\rho\frac{(x-\mu_X)(y-\mu_Y)}{\sigma_X\sigma_Y} + \frac{(y-\mu_Y)^2}{\sigma_Y^2}\right]\right)} \quad (3.24)$$

It is left as an exercise for the reader to verify that the marginal and conditional densities are as follows:

$$f_X(x) = \frac{1}{\sigma_X\sqrt{2\pi}} e^{-\frac{1}{2}\left[\frac{(x-\mu_X)}{\sigma_X}\right]^2} \qquad (3.25)$$

$$f_Y(y) = \frac{1}{\sigma_Y\sqrt{2\pi}} e^{-\frac{1}{2}\left[\frac{(y-\mu_Y)}{\sigma_Y}\right]^2} \qquad (3.26)$$

$$f_{Y|X}(x,y) = \frac{1}{\sigma_Y\sqrt{2\pi(1-\rho^2)}} e^{-\frac{1}{2}\left(\frac{y-\left\{\mu_Y+\rho\frac{\sigma_Y}{\sigma_X}(x-\mu_X)\right\}}{\sigma_Y\sqrt{(1-\rho^2)}}\right)^2} \qquad (3.27)$$

The impact of correlation on the shape of the density function (Equation 3.24) is illustrated in Figure 3.5.

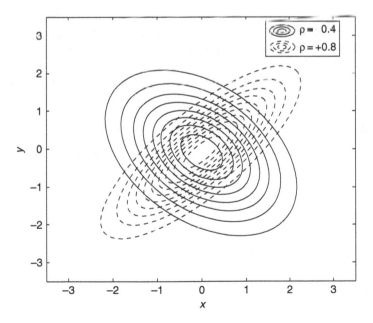

Figure 3.5 Examples of positive and negative correlation showing probability contours of the bivariate normal density.

3.3   Functions of more than one variable

In much of what will be discussed in later chapters we will be interested in functions of two or more random variables and the statistics of this function. Given the random variables X and Y and a function $g(x,y)$, then $g(X, Y) = Z$ is also a random variable. The statistics of Z can be expressed in terms of the function $g(x,y)$ and the joint statistics of X and Y. Now, for a given value z, the region(s) of the x–y plane for which $g(x,y) \le z$ is denoted by D_z, so

$$\{Z \le z\} = \{g(X, Y) \le z\} = \{(X, Y) \in D_z\}$$

Thence,

$$F_Z(z) = P\{Z \le z\} = P\{(X, Y) \in D_z\} = \iint\limits_{D_z} f_{XY}(x,y)dxdy \tag{3.28}$$

Therefore, to find $F_Z(z)$ requires us to determine the region D_z for every z and to evaluate the integral in Equation (3.28). The density can be determined by differentiating the distribution function, or directly, as

$$f_Z(z)dz = P\{z < Z \le z + dz\} = \iint\limits_{\Delta D_z} f_{XY}(x,y)dxdy \tag{3.29}$$

While this procedure may sound straightforward, even if somewhat involved, it can be severely complicated by the fact that the region D_z does not have to be connected. In some applications a significant part of the work can be identifying all the areas that make up D_z. A few examples of simple forms for $g(x,y)$ are described below because they will reappear in later chapters.

Example 3.8. $Z = g(X, Y) = X + Y.$

Solution. The region D_z such that $x + y \le z$ is the region to the left of the line $x + y = z$ (shaded in Figure 3.6).

For fixed $y, x = z - y$; thus, integrating over all values of y, the corresponding values of x are $z - y$:

$$F_Z(z) = \int\limits_{-\infty}^{\infty} \int\limits_{-\infty}^{z-y} f_{XY}(x,y)dxdy \tag{3.30}$$

Differentiating with respect to x yields

$$f_Z(z) = \int\limits_{-\infty}^{\infty} f_{XY}(z - y, y)dy \tag{3.31}$$

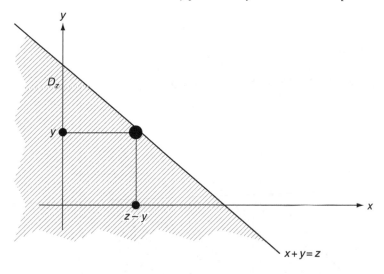

Figure 3.6 Illustration of the variables for the transformation in Example 3.8.

If X and Y are independent, then their joint density function is the product of their marginal density functions, and so:

$$f_Z(z) = \int_{-\infty}^{\infty} f_X(z-y) f_{Y(y)} \, dy \tag{3.32}$$

This is the convolution of the functions $f_X(x)$ and $f_Y(y)$. Equation (3.32) plays an important role in the simpler versions of reliability theory.

Example 3.9. $Z = g(X, Y) = X - Y$.

Solution. Replacing Y with $-Y$ in the previous example we obtain

$$f_Z(z) = \int_{-\infty}^{\infty} f_X(z+y) f_{Y(y)} \, dy \tag{3.33}$$

Example 3.10. $Z = g(X, Y) = X/Y$.

Solution. The region D_z for which $x/y \leq z$ is the shaded portion in Figure 3.7.

Integrating over suitable strips we find that

$$F_Z(z) = \int_{0}^{\infty} \int_{-\infty}^{yz} f_{XY}(x,y) \, dx \, dy + \int_{-\infty}^{0} \int_{yz}^{\infty} f_{XY}(x,y) \, dx \, dy \tag{3.34}$$

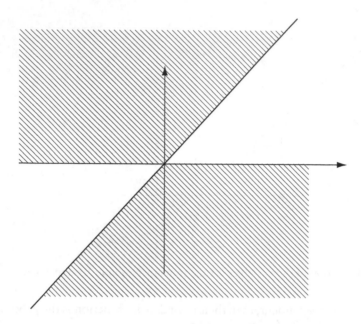

Figure 3.7 Illustrating the regions used in Example 3.10.

and

$$f_Z(z) = \int\limits_{-\infty}^{\infty} |y| f_{XY}(zy, y) dy \tag{3.35}$$

The function $g(x, y)$ defined in this case is used to define the safety factor of a system that has a random strength X and random loading Y. In passing, it is noted that if both X and Y follow a normal distribution then Z has a Cauchy density and thus no moments.

Example 3.11. $Z = g(X, Y) = XY$.

Solution. The region D_z for which $xy \leq z$ is the shaded portion in Figure 3.8. Integrating over suitable strips we find that

$$F_Z(z) = \int\limits_{0}^{\infty} \int\limits_{-\infty}^{z/x} f_{XY}(x, y) dy dx + \int\limits_{-\infty}^{0} \int\limits_{z/x}^{\infty} f_{XY}(x, y) dy dx \tag{3.36}$$

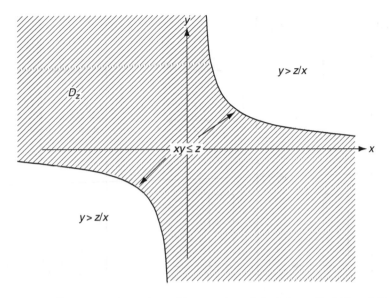

Figure 3.8 Illustrating the region of integration of the joint pdf of X and Y used in Example 3.11.

and

$$f_Z(z) = \int_{-\infty}^{\infty} \frac{1}{|y|} f_{XY}\left(\frac{z}{y}, y\right) dy = \int_{-\infty}^{\infty} \frac{1}{|x|} f_{XY}\left(x, \frac{z}{x}\right) dx \qquad (3.37)$$

The function $g(x,y)$ defined in this case is used to describe the total cost of a system that has a random demand X and random cost per unit of demand Y.

Example 3.12. $Z = g(X, Y) = X + Y$, where X and Y are independent and obey uniform distributions over intervals $a < x < b$ and $c < y < d$.

Solution. It follows from Equation (3.32) that the convolution of two rectangular density functions is a trapezoidal density. If the random variables X and Y are uniform in the intervals (a, b) and (c, d), respectively, then $Z = X + Y$ has a trapezoidal density. If, in addition, $b - a = d - c$, then the density of Z is triangular. This is illustrated in Figure 3.9.

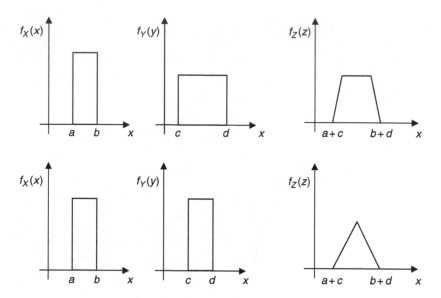

Figure 3.9 Transformation of the sum of two uniformly distributed random variables to a trapezoidal or triangular distribution.

Example 3.13. $Z=g(X, Y)=X+Y$, where X and Y are independent $N(0, 1)$ variables.

Solution. From Equation (3.32), it may be determined that Z has density given by

$$f_Z(z) = \int\limits_{-\infty}^{\infty} f_X(z-y)f_{Y(y)}dy = \int\limits_{-\infty}^{\infty} \frac{1}{\sqrt{2\pi}}e^{-\frac{1}{2}(z-y)^2}\frac{1}{\sqrt{2\pi}}e^{-\frac{1}{2}y^2}dy$$

$$= \frac{1}{2\sqrt{\pi}}e^{-\left(\frac{z}{2}\right)^2}\int\limits_{-\infty}^{\infty} \frac{1}{\sqrt{2\pi}}e^{-\frac{1}{2}u^2}du \qquad (3.38)$$

$$= \frac{1}{2\sqrt{\pi}}e^{-\left(\frac{z}{2}\right)^2}$$

where the substitution $u = (x-z/2)\sqrt{2}$ has been used.

Example 3.14. $Z=g(X, Y)=XY$, where X and Y are independent exponential variables with parameters a and b, respectively.

Solution. From Equation (3.36), it may be determined that Z has distribution given by

$$F_Z(z) = \int\limits_0^\infty \int\limits_{-\infty}^{z/x} f_{XY}(x,y)\,dy\,dx + \int\limits_{-\infty}^0 \int\limits_{z/x}^\infty f_{XY}(x,y)\,dy\,dx$$

$$= \int\limits_{-\infty}^z \int\limits_{-\infty}^\infty \frac{1}{|y|} f_{XY}\left(\frac{t}{y}, y\right) dy\,dt \qquad (3.39)$$

$$= \int\limits_0^z \int\limits_0^\infty \frac{ab}{y} e^{-(by + (at/y))}\,dy\,dt$$

$$= 1 - 2\sqrt{abz}\,K_1(2\sqrt{abz})$$

where K_1 is the modified Bessel function of order one. This form of distribution function has found application is describing the intensity of acoustic and radar waves reflected from irregular surfaces (see e.g., Jakeman 1980 and Shankar et al. 2000).

Example 3.15. $Z = g(X, Y) = \sqrt{(X^2 + Y^2)}$, where X and Y are independent $N(0, \sigma)$ variables.

Solution. If

$$f_{XY}(x,y) = \frac{1}{2\pi\sigma^2} e^{-\left(\frac{x^2 + y^2}{2\sigma^2}\right)} \qquad (3.40)$$

then, transforming to polar coordinates (r, θ) gives

$$F_Z(z) = \frac{1}{\sigma^2} \int\limits_0^z re^{-\left(\frac{r^2}{2\sigma^2}\right)}\,dr = 1 - e^{-\frac{z^2}{2\sigma^2}} \qquad z > 0 \qquad (3.41)$$

and hence

$$f_Z(z) = \frac{z}{\sigma^2} e^{-\left(\frac{z^2}{2\sigma^2}\right)} \qquad z > 0 \qquad (3.42)$$

Thus, Z has a Rayleigh density.

Example 3.16. Consider the sine wave $X\cos(\omega t) + Y\sin(\omega t) = Z\cos(\omega t + \theta)$. Since $Z = \sqrt{(X^2 + Y^2)}$, from the previous example we know that if X and Y are $N(0, \sigma)$ then the density of Z is Rayleigh. The significance of this is that, if one considers the sea surface to be composed of sinusoidal waves with random amplitudes following a normal distribution, then the resulting wave heights follow a Rayleigh distribution. This has been found to be a reasonable approximation for swell waves, which tend to have a smaller range of amplitudes and phases than wind waves.

Example 3.17. Maximum and minimum of two or more variables. In the general case of n variables, $X_1, X_2, \ldots, X_n$ we define $Y = \max(X_1, X_2, \ldots, X_n)$. The distribution function of Y will be the joint probability that each variable X_i is less than or equal to y. In other words,

$$P(Y \leq y) = P(X_1 \leq y, X_2 \leq y, \ldots, X_n \leq y) = F_{X_1 X_2 \ldots X_n}(y, y, \ldots, y) \qquad (3.43)$$

If the X_i are mutually independent then the probability is equal to the product of the individual probabilities, so that

$$F_Y(y) = \prod_{i=1}^{n} F_{X_i}(y) = (F_X(y))^n \qquad (3.44)$$

if the n variables have a common distribution function F_X. Following a similar line of argument, the variable $Z = \min(X_1, X_2, \ldots, X_n)$ may be shown to have the distribution function

$$F_Z(z) = 1 - \prod_{i=1}^{n} [1 - F_{X_i}(z)] = 1 - [1 - F_X(y)]^n \qquad (3.45)$$

when the X_i are identically distributed and independent. The corresponding density function can be obtained by differentiation in a straightforward manner:

$$f_Z(z) = \frac{dF_Z(z)}{dz} = n[1 - F_X(z)]^{n-1} f_X(z) \qquad (3.46)$$

With due attention to the requirement for independence these results can be used to 'scale up' or 'scale down' observations for different periods. For example, suppose we have a time sequence of daily observations of rainfall, with a known distribution function. Equation (3.44) may be used to determine the distribution of the annual maxima *but only* if the daily observations are independent. In a similar fashion, if one knows the distribution of the annual maxima, then the maxima over a period of $(1/n)$th of a year may be found by taking the nth root. Again, caution should be exercised in the choice of n due to the inherent assumption of independence.

3.4 Central limit theorem

The central limit theorem is an important result from the theory of probability. This states that if the random variables X_i are independent, then the density of their sum tends to a normal distribution as $n \to \infty$. This is a powerful result, which is misused typically through assuming that variables are independent when in fact they are not. There are other caveats regarding the

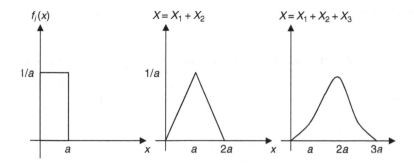

Figure 3.10 Progression of the sum of uniformly distributed random variables, illustrating the tendency towards normality as the sum is performed over a larger number of variables.

distributions of the X_i, essentially that for the theorem to hold there must not be too many variables with very narrow distributions in the total of n variables. Another situation where the assumption that variables are normal is not helpful is in the description of extreme values of a distribution. These are discussed in more detail in Chapter 4. We conclude this section with a simple illustration ('proof by pictures') of how the central limit theorem works.

Example 3.18. Let the random variables $X_i (i = 1, 2, \ldots, n)$, be independent and identically distributed with uniform density of $1/a$ in the interval $[0, a]$. Calculate the density for the sum of 2 and 3 variables (i.e., for $n = 2$ and $n = 3$).

Solution. We have $\mu_i = a$ and $\sigma_i^2 = a^2/12$. If $n = 2$, then $\mu = 2a$ and $\sigma^2 = a^2/6$. The density function in this case is triangular, obtained by convolving a rectangle with itself. If $n = 3$, then $\mu = 3a/2$ and $\sigma^2 = a^2/4$, and the density function consists of three parabolic pieces obtained by convolving the rectangular distribution with a triangle. Figure 3.10 illustrates the progression as more variables are added to the sum.

3.5 Characteristic and moment generating functions

A function that is sometimes useful in the analytical manipulation of distributions is the *characteristic function of a random variable*. For a random variable x this is defined as

$$\Phi(\omega) = E(e^{i\omega x}) = \int_{-\infty}^{\infty} f(x)e^{i\omega x}\,dx \tag{3.47}$$

which is related to the Fourier transform of the density function. The density function of a random variable can be recovered from its characteristic function via the inverse Fourier transform.

Example 3.19. Let X be $N(0,\sigma)$ and $Y=aX^2$. Find the density function of Y.

Solution. Substituting the normal distribution for $f(x)$ in Equation (3.47), and noting that the integrand is an even function, we find

$$\Phi_Y(\omega) = \int_{-\infty}^{\infty} f(x)e^{ia\omega x^2}\,dx = \frac{2}{\sigma\sqrt{2\pi}} \int_0^{\infty} e^{-x^2/2\sigma^2} e^{ia\omega x^2}\,dx \tag{3.48}$$

As x increases from 0 to ∞ the transformation $y=ax^2$ is one-to-one, that is, for each value of x there is only one value of y. Now, $dy = 2ax\,dx = 2\sqrt{(ay)}\,dx$, so

$$\Phi_Y(\omega) = \frac{2}{\sigma\sqrt{2\pi}} \int_0^{\infty} \frac{e^{-y/2a\sigma^2}\,e^{i\omega y}}{2\sqrt{ay}}\,dy \tag{3.49}$$

and thus, from the definition of the characteristic function,

$$f_Y(y) = \frac{e^{-y/2a\sigma^2}}{\sigma\sqrt{2a\pi y}} \tag{3.50}$$

Another function, related to the characteristic function, which can be helpful in determining the moments of a random variable, is the *moment generating function*, $M(s)$. For a random variable X, this is defined by

$$M(s) = E(e^{sx}) = \int_{-\infty}^{\infty} f(x)e^{sx}\,dx \tag{3.51}$$

which is related to the Laplace transform of the density function. Note that $M(i\omega) = \Phi(\omega)$. If we differentiate Equation (3.51) with respect to s we have:

$$\frac{dM(s)}{ds} = \int_{-\infty}^{\infty} sf(x)e^{sx}\,dx = E(xe^{sx}) \tag{3.52}$$

Repeating n times we find

$$\frac{d^n M(s)}{ds^n} = E(x^n e^{sx})$$

$$\Rightarrow \frac{d^n M(0)}{ds^n} = E(x^n) = m_n \tag{3.53}$$

Example 3.20. If X has an exponential density with parameter a, find its moment generating function and use this to determine the mean and variance.

Solution. From Equation (3.51),

$$M(s) = \int_0^\infty ae^{-ax}e^{sx}dx = \frac{a}{a-s} \tag{3.54}$$

also,

$$\frac{dM(0)}{ds} = \frac{1}{a}$$
$$\frac{d^2M(0)}{ds^2} = \frac{2}{a^2} \tag{3.55}$$

so

$$E(X) = \frac{1}{a} \quad E(X^2) = \frac{2}{a^2} \quad \text{and} \quad \sigma^2 = E(X^2) - E(X)^2 = \frac{1}{a^2} \tag{3.56}$$

Example 3.21. If Y has a normal density with a mean and standard deviation μ_Y and σ_Y, respectively, determine the mean and variance of X, where $Y = \ln(X)$.

Solution. The distribution that X obeys is known as a log–normal distribution. Using the transformation in Equation (3.1) the probability density function of X is

$$f_X(x) = \frac{1}{x\sigma_{\ln(X)}\sqrt{2\pi}} \exp\left\{-\frac{1}{2}\left[\frac{\ln(x) - \mu_{\ln(X)}}{\sigma_{\ln(X)}}\right]^2\right\} \tag{3.57}$$

for $0 \leq x < \infty$. Now, either from the moment generating function, or the definition of the nth moments, the mean and variance of X may be determined as follows:

$$E(X^r) = \int_0^\infty x^r f_X(x)dx = \frac{1}{\sigma_{\ln(X)}\sqrt{2\pi}} \int_{-\infty}^\infty e^{sy} \exp\left\{-\frac{1}{2}\left[\frac{y - \mu_{\ln(X)}}{\sigma_{\ln(X)}}\right]^2\right\} dy$$

$$= \exp\left(s\mu_Y + 0.5s^2\sigma_Y^2\right) \tag{3.58}$$

But the right-hand side of the above equation is exactly the definition of the moment generating function for Y! Therefore

$$E(X^r) = M_Y(s)$$

$$E(X) = \exp\left(\mu_{\ln(X)} + 0.5\sigma_{\ln(X)}^2\right) \equiv \mu_X$$

$$E(X^2) = \exp\left(2\left[\mu_{\ln(X)} + \sigma_{\ln(X)}^2\right]\right) \tag{3.59}$$

$$Var(X) = E(X^2) - E(X)^2 = \mu_X^2\left\{\sigma_{\ln(X)}^2 - 1\right\}$$

Note also from the above that $\ln(\mu_X) = \mu_{\ln(X)} + 0.5\sigma_{\ln(X)}{}^2$, and

$$\left(\frac{\sigma_X}{\mu_X}\right)^2 + 1 = \sigma_{\ln(X)}^2 \tag{3.60}$$

3.6 Transformation of moments

In the previous sections it has been demonstrated how the probability density function of a function of a random variable can be determined from a knowledge of the probability density function of the original variable. This is not always straightforward, and for nonlinear transformations can be particularly difficult. The same comments apply to the case of transformations involving two or more random variables. However, it may be possible to determine the moments of a transformed variable more easily, and these can give useful practical information about the behaviour of the transformed variable.

The general expression for the rth moment of a function $Y = g(X_1, X_2, \ldots, X_n)$ of n random variables $X_1, X_2, \ldots, X_n$ may be written as:

$$E(Y^r) = \int_{-\infty}^{\infty} \cdots \int_{-\infty}^{\infty} Y(x_1, x_2, \ldots, x_n) f_{X_1 X_2 \ldots X_n}(x_1, x_2, \ldots, x_n) dx_1 dx_2 \ldots dx_n$$

$$\tag{3.61}$$

A summary of some important special cases is given below without workings.

$$Y = \sum_{i=1}^{n} a_i X_i \qquad E(Y) = \sum_{i=1}^{n} a_i \mu_{X_i} \qquad Var(Y) = \sum_{i=1}^{n} a_i^2 Var(X_i)$$

$$+ \sum_{i=1}^{n} \sum_{j=1}^{n} a_i a_j Cov(X_i, X_j)$$

$$Y = X_1 X_2 \qquad E(Y) = \mu_{X_1}\mu_{X_2} + \rho\sigma_{X_1}\sigma_{X_2} \qquad \mathrm{Var}(Y) = (1+\rho^2)\big[(\mu_{X_1}\sigma_{X_1})^2$$
$$+ (\mu_{X_2}\sigma_{X_2})^2 + (\sigma_{X_1}\sigma_{X_2})^2\big]$$

$$Y = \prod_{i=1}^{n} X_i \qquad X_i \text{ independent, } E(Y) = \prod_{i=1}^{n}\mu_{X_i} \quad \mathrm{Var}(Y) = \prod_{i=1}^{n}\mu_{X_i}^2 - \left(\prod_{i=1}^{n}\mu_{X_i}\right)^2$$

$$Y = \sqrt{X} \qquad E(Y) = \left(\mu_X^2 - \frac{\sigma_X^2}{2}\right)^{1/4} \qquad \mathrm{Var}(Y) = \mu_X - \left(\mu_X^2 - \frac{\sigma_X^2}{2}\right)^{1/2}$$

$$Y = aX^2 + bX + c \quad E(Y) = a(\mu_X^2 + \sigma_X^2) + b\mu_X + c \qquad \mathrm{Var}(Y) = \sigma_X^2(2a\mu_X + b)^2$$
$$+ 2a^2\sigma_X^4$$

$$(3.62)$$

where ρ is the correlation coefficient.

Example 3.22. Hydrographic surveying of river and sea beds measured by echo-sounder is subject to two independent sources of error: movements of the vessel and electronic noise in the processing circuitry. If these two sources are described by normal variates X and Y with mean values μ_X and μ_Y, and variances σ_X^2 and σ_Y^2, respectively, find the mean and variance of the overall error $Z = X + Y$.

Solution. We will use the moment generating function to solve this problem. The moment generating function for X is given by:

$$M_X(s) = E(e^{sx}) = \int_{-\infty}^{\infty} \frac{1}{\sigma_X\sqrt{2\pi}} e^{-\frac{(x-\mu_X)^2}{2\sigma_X^2}} e^{sx} dx$$

$$= \frac{e^{s\mu_X}}{\sigma_X\sqrt{2\pi}} \int_{-\infty}^{\infty} e^{-\frac{[(x-\mu_X-\sigma_X^2 s)^2 - \sigma_X^4 s^2]}{2\sigma_X^2}} dx$$

$$= \frac{e^{s\mu_X} e^{\sigma_X^2 s^2/2}}{\sigma_X\sqrt{2\pi}} \int_{-\infty}^{\infty} e^{-\frac{(x-\mu_X-\sigma_X^2 s)^2}{2\sigma_X^2}} dx$$

$$= e^{(2\mu_X s + \sigma_X^2 s^2)/2}$$

The moment generating function for Y will have an analogous form. Now, as X and Y are independent

$$M_Z(s) = E(e^{sZ}) = E(e^{s(X+Y)}) = E(e^{sX}e^{sY}) = E(e^{sX})E(e^{sY}) = M_X(s)M_Y(s)$$

Thus, the moment generating function of Z is:

$$M_Z(s) = e^{(2\mu_X s + \sigma_X^2 s^2)/2} e^{(2\mu_Y s + \sigma_Y^2 s^2)/2} = e^{[(\mu_X + \mu_Y)s + (\sigma_X^2 + \sigma_Y^2)s^2/2]}$$

and

$$\frac{dM_Z(0)}{ds} = \mu_X + \mu_Y$$

$$\frac{d^2 M_Z(0)}{ds^2} = (\mu_X + \mu_Y)^2 + \sigma_X^2 + \sigma_Y^2$$

$$\Rightarrow \mathrm{Var}(Z) = \sigma_X^2 + \sigma_Y^2$$

Therefore, the mean overall error in the measurements is the sum of the means of the individual errors; and its variance is the sum of the variances.

The mean and variance of general functions is often difficult to obtain due to the complexity in evaluating the transformation integrals. Estimates of the mean and variance can be approximated in the following manner. First, expand the function $Y = g(X_1, X_2, \ldots, X_n)$ in a Taylor series about the point defined by the vector of the means $(\mu_1, \mu_2, \ldots, \mu_n)$. Now truncate the series expansion at linear terms. The first-order estimates of the mean and variance of Y are:

$$E(Y) \approx Y(\mu_1, \mu_2, \ldots, \mu_n)$$

$$\mathrm{Var}(Y) \approx \sum_i^n \sum_j^n c_i c_j \mathrm{Cov}(X_i, X_j) \tag{3.63}$$

where

$$c_i \equiv \left. \frac{\partial Y}{\partial X_i} \right|_{\mu_1, \mu_2, \ldots, \mu_n}.$$

If the X_i are independent, then $\mathrm{Cov}(X_i, X_j) = 0$ if $i \neq j$ and $\mathrm{Cov}(X_i, X_j) = \mathrm{Var}(X_i)$ if $i = j$. Corresponding expressions can be obtained by truncating the Taylor series expansion at higher orders. As an example, the second-order approximations for a function of two variables are

$$E(Y) \approx Y(\mu_1, \mu_2, \ldots, \mu_n) + \frac{1}{2} \frac{\partial^2 Y}{\partial x_1^2} \mathrm{Var}(X_1) + \frac{1}{2} \frac{\partial^2 Y}{\partial x_2^2} \mathrm{Var}(X_2) + \frac{\partial^2 Y}{\partial x_1 \partial x_2} \mathrm{Cov}(X_1, X_2)$$

$$\mathrm{Var}(Y) \approx \left(\frac{\partial Y}{\partial x_1} \right)^2 \mathrm{Var}(X_1) + \left(\frac{\partial Y}{\partial x_2} \right)^2 \mathrm{Var}(X_2) + 2 \left(\frac{\partial Y}{\partial x_1} \frac{\partial Y}{\partial x_2} \right) \mathrm{Cov}(X_1, X_2) \tag{3.64}$$

where the derivatives are evaluated at the mean values of the variables.

3.7 Estimating parameters

In most practical reliability problems the exact distribution of random variables is not known. Rather, a set of observations will have been gathered. The standard approach is then to select a distribution with a particular analytical form to fit the data by suitable adjustment of the distribution parameters. (A distribution with arbitrary parameters, such as the normal distribution with mean μ and standard deviation σ, is considered a 'family of distributions'. The process of fitting the normal distribution to the observations will result in specific values of μ and of σ that make the normal distribution fit the observations most closely.) Once this has been done, the analytical function may be used to interpolate between measured data points, and more often to extrapolate beyond the data to determine extreme values for particular quantiles. Without an analytical expression for the distribution function it is much more difficult to extrapolate to extreme values, but fitting a particular functional form to the data imposes a constraint on the measurements that may not be justified. For this reason the selection of distribution functions, or *statistical model*, is an extremely important process, and the associated uncertainties should be understood. The choice of model may be influenced by knowledge of the physical processes important in the situation being studied. Past experience may also be valuable; if similar observations have led to results conforming to a particular distribution then this indicates a useful starting point. In the absence of any guide, exploratory data analysis is required to investigate the characteristics of the empirical distribution function, which in turn may suggest plausible families of distributions. In any fitting process, the observations are considered to be a sample of a larger population the characteristics of which are being estimated, and the properties of the sample are used to infer the properties of the whole population. The parameter values determined by fitting to a sample are termed an *estimate*.

Before describing various methods for estimating parameters of probability distributions, three important concepts are introduced. These are *consistency*, *bias* and *efficiency*. Parameter estimates are said to be *consistent* if they converge in probability as the sample size increases. Estimates are *unbiased* if an average is taken of the estimates of the parameters from a large number of samples of the same size and the average converges to the true value of the parameter, that is, $E(\theta') = \theta$, where θ' is the estimator of the parameter θ. Estimates that do not have this property are *biased*. An unbiased estimator θ' may still deviate widely from θ. A desirable property of an estimator is that its variance should be as small as possible (thereby reducing the uncertainty in any estimate); that is, it should be *efficient*.

Different *estimates* are considered to be different values of a function, the *estimator*. Any estimate will have an associated uncertainty, which is measured by confidence limits. In general, an estimate θ' is given by some function, f, of a random sample from the population, $X = \{X_1, X_2, \ldots, X_n\}$. For large samples, many estimates may be approximated as normally distributed, and confidence bounds are defined in the form

$$E[f(X)] \pm z_c \sqrt{Var[f(X)]} \qquad (3.65)$$

where z_c is a percentage point from the standard normal distribution reflecting the degree of belief in the limits. For example, the 90% confidence limits are based on the 5% and 95% points of the standard normal distribution. The variability in parameter estimates is measured by the standard error (s.e.), given by the square root of the variance of the estimator.

3.7.1 Method of moments

The method of moments is a well-established method. It works by equating the sample moments to the analytical expressions of the moments for the chosen distribution function.

Example 3.23. The mean and standard deviation of a set of near-shore wave heights are 2 m and 0.5 m, respectively. Find the distribution parameters assuming (a) a Rayleigh distribution and (b) a gamma distribution.

Solution. (a) The Rayleigh distribution has one parameter and:

$$F(x) = 1 - \exp\left[-\frac{x^2}{2b^2}\right], \qquad 0 \le x$$

$$\text{Mean} = b\sqrt{\frac{\pi}{2}}, \qquad \text{Variance} = (2 - \pi/2)b^2 \qquad (3.66)$$

Using the expression for the mean gives $2 = b\sqrt{(\pi/2)}$, or $b = 1.60$. Note that this value of b gives a standard deviation of 1.0 m, which does not agree very well with the sample standard deviation. Using the expression for the variance to solve for b gives $b = \sqrt{[0.25/(2 - \pi/2)]} = 0.76$. In turn this gives a value for the mean of 0.95 m, which again is not a good match with the sample mean. This suggests that the Rayleigh distribution is not a good statistical model for the observations.

(b) The gamma distribution has two parameters, λ and r, both of which must be positive (see Equation 2.23). The density function is defined as:

$$f(x) = \frac{\lambda^r}{\Gamma(r)} e^{-\lambda x} x^{r-1} \qquad 0 \le x < \infty \qquad (3.67)$$

$$\text{Mean} = r/\lambda \qquad \text{Variance} = \lambda^2 r$$

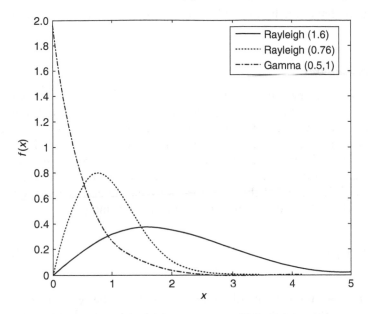

Figure 3.11 The density functions corresponding to Rayleigh(1.6), Rayleigh(0.76) and Gamma(0.5, 1).

Setting the expressions for the mean and variance equal to the sample moments yields two simultaneous equations for λ and r. Using the expression for the mean to eliminate r from the expression for the variance gives $\lambda = 1/2$, and thus $r = 1$. The functions are shown in Figure 3.11.

The extra parameter in the gamma distribution gives extra flexibility for the distribution to fit the data. The nature of the density functions is very different, and when using the method of moments it is often advantageous to use additional information to choose an appropriate family of density functions. For distributions with n parameters, n moments are required, yielding n simultaneous equations to be solved. The advantage of the method of moments is that it gives estimates that are easily obtained. The estimates are also usually consistent. However, the method is not suitable for fitting distributions that have no moments (e.g., Cauchy). Furthermore, the method can lead to biased estimates, and is not always efficient for small sample sizes (say, $n < 30$) or if higher order (third and above) moments are used.

3.7.2 *Methods of probability-weighted and L-moments*

The methods of probability-weighted moments and L-moments are variants of the method of moments that seek to avoid some of the difficulties encountered with evaluating high-order moments. Details of these methods can be found in Greenwood et al. (1979) and Hosking (1990). Although useful, these methods have not found wide application in hydraulic engineering practice.

3.7.3 Least squares

The least-squares technique is probably the default choice for many cases. It is typically based on fitting a straight line to observations, and minimising a function of the distances of the points from the line. A straight line can be written as $y = mx + c$. For fitting observations to a probability distribution, x_i would be the observed data in increasing order, and y_i would be the empirical distribution function. The least-squares principle is that the sum of the squared deviations of the distances between the points and the line, measured in the y-direction, should be minimised. Defining,

$$R^2 = \sum_{i=1}^{n} (y_i - \hat{y}_i)^2 \tag{3.68}$$

where n is the number of observations and is the estimator of y (i.e., $= mx_i + c$), the least-squares estimates of m and c are obtained by differentiating R^2 with respect to m and c, setting each derivative to zero and solving for m and c. Thus,

$$\frac{dR^2}{dm} = -2 \sum_{i=1}^{n} x_i[y_i - (mx_i + c)] = 0 \tag{3.69}$$

$$\frac{dR^2}{dc} = -2 \sum_{i=1}^{n} [y_i - (mx_i + c)] = 0 \tag{3.70}$$

Equations (3.69) and (3.70) lead to a set of simultaneous equations that have to be solved to find m and c. The equations can be cast in matrix form:

$$\begin{pmatrix} n & \sum x_i \\ \sum x_i & \sum x_i^2 \end{pmatrix} \begin{pmatrix} c \\ m \end{pmatrix} = \begin{pmatrix} \sum y_i \\ \sum x_i y_i \end{pmatrix} \tag{3.71}$$

which has the solution

$$\hat{c} = \frac{\sum x_i^2 \sum y_i - \sum x_i \sum x_i y_i}{n \sum x_i^2 - (\sum x_i)^2} \tag{3.72}$$

and

$$\hat{m} = \frac{n \sum x_i y_i \sum y_i - \sum x_i \sum y_i}{n \sum x_i^2 - (\sum x_i)^2} \tag{3.73}$$

A good measure of fit can often be gained simply by looking at a plot of the points and the fitted line. A more quantitative measure of fit is given by the value of R^2. In software packages this value is often normalised so that it lies between 0 and 1, with 1 corresponding to a perfect fit. Now, there is no reason why we should restrict ourselves to a linear function. Indeed, the analysis above can be extended to a wide range of functions, although the algebra can get very extensive. In fact, it is often more efficient to transform the variables x and y first, in order to fit a straight line to the (transformed) points.

Example 3.24. It is required to fit an exponential curve $y = ae^{bx}$ to a set of observations (x_i, y_i). Find the transformation of variables that will allow the use of straight-line fitting.

Solution. Taking logarithms of both sides of the equation yields:

$$\ln(y) = b\ln(x) + \ln(a)$$

using the properties of logarithms. But this is in the form $y = mx + c$, so by using the points $\ln(x_i)$, $\ln(y_i)$ in the straight-line fitting algorithm should give a straight line that intercepts the y-axis at $y = \ln(a)$ and has slope b.

Example 3.25. The River Vistula in Poland runs into the Bay of Gdansk. There is a benchmark station at Tczew, on the banks of the River Vistula, about 35 km upstream from the mouth of the river. Table 3.1 shows the measured annual maximum river levels in millimetres above the zero datum. Observations x_i of annual peak river level are likely to follow the following distribution:

$$F_X(x) = \exp\left[-\exp\left(-\frac{(x-a)}{b}\right)\right]$$

Estimate the parameters a and b of the distribution that best fit the observations.

Solution. First, put the x_i in order of increasing value. Then compute the empirical distribution $F_X(x_i) = i/(N+1)$. Writing $y = F_X(x)$, we find that:

$$-\frac{x}{b} + \frac{a}{b} = \ln[-\ln(y)]$$

Therefore, by plotting $\ln[-\ln(F_X(x_i))]$ against x_i should give a straight-line fit. The slope of the line will be $-1/b$ and the intercept will be a/b. Table 3.1 summarises the data and Figure 3.12 shows the resulting points and straight-line fit. Parameters a and b are found as $a = 719.8$ mm and $b = 116.1$ mm. Visually, the straight-line fit appears reasonable, particularly for those points in the middle of the range. However, the fit appears

Table 3.1 Annual maximum river levels

Year	River level (mm)
1961	566
1962	1017
1963	698
1964	927
1965	905
1966	908
1967	862
1968	715
1969	656
1970	900
1971	806
1972	772
1973	640
1974	867
1975	801
1976	738
1977	814
1978	696
1979	991
1980	981
1981	841
1982	844
1983	716
1984	507
1985	650
1986	642
1987	785
1988	730
1989	704

less good at the extremes of the range where, naturally, there are fewer observations.

Plots like Figure 3.12 are known as quantile–quantile or Q-Q plots. There are some choices to be made about the *x*-ordinate that should be used when generating the empirical distribution, and this is discussed further in Chapter 4.

3.7.4 Maximum likelihood estimation

The maximum likelihood procedure is another way of estimating the parameters of a distribution. It has several desirable properties that often make it the method of choice. The method requires the construction of a *likelihood function*, *L*, which is a function of the observed data and the chosen density function. This function is maximised, for the given dataset, with respect

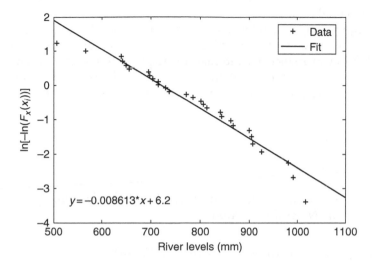

Figure 3.12 Straight-line fit to the River Vistula measurements (*Q-Q* plot) for Example 3.25.

to the chosen distribution function parameters. The principle is first illustrated for the binomial distribution. Consider repeated independent events that have a constant probability, p, of success. The probability of m successes in n independent trials is $^nC^m p^m q^{n-m}$ (see Equation 2.12). We have been given a coin that may or may not be biased, and told that in 15 tosses of the coin 9 heads and 6 tails were obtained. The likelihood function is $L = {}^nC^m p^m q^{n-m}$. We now investigate the value of the likelihood function for different values of the probability of obtaining 'heads'.

If $p = 0.4$ the probability of a sample
result such as that given would be: $^{15}C^9(0.4)^9(0.6)^6 = 0.061;$
If $p = 0.5$ the probability becomes: $^{15}C^9(0.5)^9(0.5)^6 = 0.153;$
And if $p = 0.6$ the probability becomes: $^{15}C^9(0.6)^9(0.4)^6 = 0.207.$

The use of the principle of maximum likelihood to decide between the three possibilities leads to the choice $p = 0.6$, as this is the value of p that would have made the given sample the most likely result.

The likelihood function is defined in general terms as:

$$L(\theta) = \prod_{i=1}^{n} f_X(x_i | \theta) \tag{3.74}$$

where $f_X(x)$ is a chosen density function and θ is the set of parameters for the density function. If L is not zero and the value of θ that maximises it is the same as that which maximises $\ln(L)$, it is often convenient to work with the log-likelihood function. This is particularly so when the density function has an exponential term.

The principle of maximum likelihood is equally applicable to continuous distributions. Suppose we are sampling from a normally distributed population with known variance σ^2, and that it is required to find the maximum likelihood estimator of the population mean μ on the basis of a sample of size N from the population $X_1, X_2, X_3, X_4, X_5, \ldots, X_N$. The density of each X_i is

$$f(X_i|\mu) = \frac{1}{\sigma\sqrt{2\pi}} e^{-(X_i - \mu)^2 / 2\sigma^2}$$

Assuming the trials to be independent, the likelihood function is simply the product of the N density functions:

$$L(X_1, X_2, \ldots, X_N|\mu) = (f(X_1|\mu)f(X_2|\mu), \ldots, f(X_i|\mu)$$

That is,

$$L(X_1, X_2, \ldots, X_N|\mu) = \left(\frac{1}{\sigma\sqrt{2\pi}} e^{-(X_1 - \mu)^2 / 2\sigma^2} \right) \left(\frac{1}{\sigma\sqrt{2\pi}} e^{-(X_2 - \mu)^2 / 2\sigma^2} \right)$$
$$\ldots \left(\frac{1}{\sigma\sqrt{2\pi}} e^{-(X_i - \mu)^2 / 2\sigma^2} \right)$$

Or,

$$L(X_1, X_2, \ldots X_N|\mu) = \left(\frac{1}{\sigma\sqrt{2\pi}} \right)^N e^{-\sum\limits_{1}^{N}(X_i - \mu)^2 / 2\sigma^2}$$

To minimise $L(X_1, X_2, \ldots, X_N|\mu)$ we take the logarithm, differentiate and set the result equal to zero:

$$\log(L) = N \log \left(\frac{1}{\sigma\sqrt{2\pi}} \right) - \frac{\sum\limits_{1}^{N}(X_i - \mu)^2}{2\sigma^2}$$

so

$$\frac{d\log(L)}{d\mu} = -\frac{\sum\limits_{1}^{N} 2(X_i - \mu)(-1)}{2\sigma^2} = \frac{\sum\limits_{1}^{N}(X_i - \mu)}{\sigma^2}$$

Setting this equal to zero we obtain:

$$\sum_{1}^{N} (X_i - \mu) = 0$$

$$or \quad \sum_{1}^{N} X_i - \sum_{1}^{N} \mu = 0$$

$$or \quad \sum_{1}^{N} X_i - N\mu = 0$$

which implies that

$$\mu = \frac{\sum_{1}^{N} X_i}{N}$$

Thus, for populations having a normal distribution, the sample mean is a maximum likelihood estimator of μ.

Confidence intervals for parameter estimates, θ'_i, based on the maximum likelihood approach, are determined as follows. The observed information matrix has entries $-\partial^2 \log(L)/\partial\theta_i\partial\theta_j$ evaluated at θ'. The inverse of this matrix is the estimated covariance matrix. Confidence intervals for the θ_i are given by: $\theta'_i \pm z_c\sqrt{[\mathrm{Var}(\theta'_i)]}$, where z_c is the chosen percentage point from the standard normal distribution.

3.7.5 Bayesian methods

Bayesian methods have had a renewed surge of development in recent years. They are, in essence, a generalisation of the maximum likelihood method that makes use of Bayes' theorem to modify an assumed distribution. The likelihood function is multiplied by a 'prior' density function that is estimated on the basis of observations or engineering judgement. The product is divided by a constant to ensure that the integral of the resulting density function is unity. In symbols:

$$P(x|y) = \frac{P(y|x)P(x)}{P(y)} \equiv \frac{\text{likelihood} \times \text{prior}}{\text{evidence}} \tag{3.75}$$

where 'x' is what is being estimated, 'y' is considered the evidence, and the 'prior' is an initial guess of the density function of the variable, x. $P(x|y)$ is referred to as the posterior, and is interpreted as an updated version of $P(x)$, improved by the knowledge of new evidence 'y'. Bayes' theorem gives a direct means of revising the density function of x as it changes from prior $P(x)$ before observation of y, to posterior $P(x|y)$ once y has been observed.

Bayesian methods work on a different philosophy from the 'classical' approach. In the classical approach, the parameters of a distribution are considered unknown constants that are to be determined. In the Bayesian approach, the unknown parameters are treated as random variables. Qualitative engineering experience may also be incorporated in the choice of prior distribution. New observations or evidence can then be used to refine the prior distribution through the application of Equation (3.75). Unsurprisingly, this approach has generated some controversy and debate, not least with regard to the choice of prior distribution and the probabilities attaching to new evidence. Some examples are discussed by Benjamin and Cornell (1970), and Kottegoda and Rosso (1997).

3.7.6   *Resampling techniques – confidence limits*

In cases where the function of the sample is not amenable to analytical treatment, or where there is an assumption that the estimates may be approximated as normally distributed, numerical techniques can be employed to find confidence limits. Two well-known numerical methods are the bootstrap[1] and the jack-knife[2] techniques. Both use the concept of re-sampling, that is, creating new samples from an existing sample. Bootstrap re-sampling has been available for more than 30 years (Efron 1979), while jack-knife re-sampling is an even older technique, and was introduced by Quenouille (1949). Both are quite computationally intensive, and have recently become more widely used with advances in computing power. Further details of re-sampling methods are described by Efron and Tibshirani (1993). A brief outline of the basic jack-knife and bootstrap procedures are given here.

Suppose a sample $\mathbf{X} = (X_1, X_2, \ldots, X_n)$ of n observations is to be used to estimate a population parameter, θ. The estimator of θ is defined as $\widehat{\theta} = f(\mathbf{X})$. The jack-knife employs samples, $\mathbf{X}_i$, that leave out one observation at a time, and which are of the form:

$$\mathbf{X}_i = (x_1, x_2, \cdots, x_{i-1}, \cdots, x_{i+1}, \cdots, x_n) \tag{3.76}$$

with $i = 1, 2, \ldots, n$. These are the jack-knife samples. For each jack-knife sample a value of the statistic in question is obtained. These are known as the 'jack-knife estimators', and are denoted by $\widehat{\theta}_i = f(\mathbf{X}_i)$. Since the process is repeated n times, there are n estimators. Now, if the function f is taken to

---

1   The term 'bootstrapping' is believed to originate from Rudolph Eric Raspe's eighteenth-century tale *The Surprising Adventures of Baron Munchausen*, in which the hero escapes from a lake, pulling himself upwards by tugging on his bootstraps.
2   'Jack-knife' alludes to the everyday usefulness of this statistical tool to the statistician.

be the mean, we have $\hat{\theta} = \sum_{i=1}^{n} x_i/n = \hat{x}$ then each jack-knife estimator of the mean will be of the form

$$\hat{\theta}_i = \sum_{j=1}^{n} \frac{x_j}{(n-1)}(1-\delta_{ij}),$$

Defining

$$\hat{\theta}_i^* \equiv n\hat{\theta} - (n-1)\hat{\theta}_i$$

the jack-knife estimate of θ is given by the average

$$\hat{\theta}^* = \frac{1}{n} \sum_{i=1}^{n} \hat{\theta}_i^*$$

and the jack-knife estimate of the variance of $\hat{\theta}$ is given by

$$\mathrm{Var}_{JK}\left(\hat{\theta}\right) = \frac{1}{n(n-1)} \sum_{i=1}^{n} (x_i - \hat{x})^2 \tag{3.77}$$

The standard error of the jack-knife estimate, $s.e._{JK}$, is given by the positive square root of Equation (3.77) and an approximate $100(1 - \beta)\%$ confidence interval for θ is given by

$$\left[\hat{\theta}^* - t_{1-\beta/2,(n-1)} s.e._{JK}\left(\hat{\theta}\right), \hat{\theta}^* + t_{1-\beta/2,(n-1)} s.e._{JK}\left(\hat{\theta}\right)\right] \tag{3.78}$$

where $t_{1-\beta/2,(n-1)}$ denotes the $(1 - \beta/2)$ quantile of a Student's t distribution with $(n-1)$ degrees of freedom.

It is well known that the jack-knife works well for linear statistics (e.g., the mean), but can fail to give accurate estimation for non-smooth (e.g., median) and nonlinear (e.g., correlation coefficient) statistics (Efron & Tibshirani 1993).

A dataset of size n has $2^n - 1$ non-empty subsets; however, jack-knife re-sampling uses only n of them. Thus, it can be seen that jack-knife re-sampling may be refined by obtaining estimates from more than n subsets, which is the main motivation for the bootstrap method, described by Efron (1979). The idea behind bootstrap re-sampling is to randomly sample the dataset a very large number of times. The assumption is that each data point is a valid member of the total sample, and that at any time there is an equal probability of 'picking' any of the data values in the original sample. Thus, new and equally valid re-sampled datasets can be created by picking from these data at random. Bootstrap re-sampling is achieved by randomly selecting n data points, with replacement, from the original observed random sample. Therefore, it is possible that in any sample of n data points some of the original data points can appear twice or more, and some of the original data points may not appear at all. All bootstrap replicates (samples) have the same length as the original samples and each of the bootstrap replicates can provide an estimator $\hat{\theta}_i^*$. The spread in the estimators formed from these re-sampled datasets then provides information on the stability of the estimator with respect to different possible outcomes represented by the bootstrap replicates. However, resampling with replacement may lead to unusual bootstrap samples. Therefore, a large number of replicates are generally recommended. Confidence intervals may be found directly from the estimators derived from the bootstrap replicates by rank ordering the estimators and selecting the appropriate values corresponding to the chosen percentage point. For example, with 1000 bootstrap replicates, the 90% confidence interval of the mean will be defined by the 50th and 950th rank-ordered values of the mean computed from the bootstrap replicates. Re-sampling methods are discussed further in Chapter 4.

3.8   Stochastic processes

In reliability analysis one often has to deal with quantities that vary as a function of time (e.g., rainfall, river flows and wave heights). To describe the time variation in reliability requires the concept of a stochastic process. The definition is analogous to that of a random variable. Formally, a stochastic process $X(t)$ is a rule for assigning to every outcome of an experiment, a, a function $X(t,a)$, that is, a stochastic process can be considered to be a family (or ensemble) of time-dependent functions that are also dependent upon the parameter a or, equivalently, a function of t and a. If a is fixed ($a = a_0$, say) then $X(t, a_0)$ is a realisation of the stochastic process. If t is fixed ($t = T$, say) and a is variable then $X(T, a)$ is a random variable equal to the state of the given process at time $t = T$. If both t and a are fixed then $X(t, a)$ is a number. Figure 3.13 illustrates this relationship.

For a specific t, $X(t, a)$ is a random variable with distribution

$$F(x,t) = P[X(t) \leq x] \tag{3.79}$$

This function depends on t, and it equals the probability of the event $\{X(t) \leq x\}$ consisting of all outcomes a_i such that, at the specific time t, the samples $X(t, a_i)$ of the given process do not exceed the number x. As an example, consider the weather forecasts produced by national forecasting offices. Nowadays, many of these base their presented forecast on an analysis of an ensemble of forecasts (typically 20–30) made with slightly different initial conditions. If we are interested in the rainfall rate at 12 noon

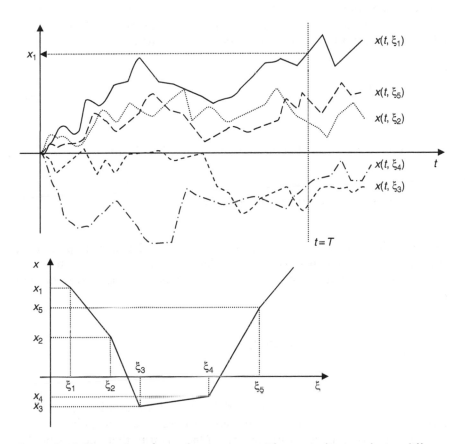

Figure 3.13 Illustration of a stochastic process. The upper diagram depicts different realisations of the process *x*, which is a function of *t* and parameter ξ. Fixing *t* = *T* we obtain *x* as a function of the parameter ξ, as shown in the lower diagram. If the value of the parameter is now fixed, we retrieve a number.

tomorrow at a particular site, then the ensemble forecasts will provide 20–30 different answers. We can rank these answers in increasing order to create a cumulative probability distribution, just as for random variables. As we move from one member of the ensemble to another we are looking at the dependence of the answer on the parameter *a* in the context of the above discussion. If we select one member of the ensemble, we are in effect fixing the parameter *a* in the above discussion, and the result is a single number. In practice, a system as complex as a weather forecasting model depends on a huge number of parameters and predicts quantities as a function of space as well as time. The same concepts of a stochastic variable apply, but the notation and manipulations become more involved.

For reasons that will become clear shortly, the function $F(x,t)$ is some-times known as the 'first-order distribution' of the process $X(t)$. In analogy with random variables, its first-order density is defined as

$$f(x,t) = \frac{\partial F(x,t)}{\partial x} \tag{3.80}$$

The second-order distribution of $X(t)$ is defined as the joint distribution

$$F(x_1, x_2; t_1, t_2) = P[X(t_1) \leq x_1, X(t_2) \leq x_2] \tag{3.81}$$

of the random variables $X(t_1)$ and $X(t_2)$. In any analytical manipulations, x_1, x_2, t_1 and t_2 are considered independent, with t_1 and t_2 describing the corresponding evolution of X in 'parallel universes'. The corresponding density is

$$f(x_1, x_2; t_1, t_2) = \frac{\partial^2 F(x_1, x_2; t_1, t_2)}{\partial x_1 \partial x_2} \tag{3.82}$$

Higher-order distributions can be defined, and are required to charac-terise a stochastic process fully. For our purposes, we need only consider the first- and second-order distributions. One quantity that is of particular importance is the autocorrelation function, $R(t_1, t_2)$, which is defined as the expected value of the product $X(t_1)X(t_2)$:

$$R(t_1, t_2) = E[X(t_1)X(t_2)] = \int_{-\infty}^{\infty} \int_{-\infty}^{\infty} x_1 x_2 f(x_1, x_2; t_1, t_2) dx_1 dx_2 \tag{3.83}$$

The autocorrelation function describes how quickly the process varies in time (i.e., becomes decoupled from earlier values) on average. Setting $t_1 = t_2$ gives the second moment, or average power, of the process $X(t)$:

$$R(t,t) = E[X^2(t)] \tag{3.84}$$

Example 3.26. Find the autocorrelation function of the process $X(t) = A\cos(\omega t + \varphi)$, where φ is uniform in the interval $(-\pi, \pi)$. This is a linear wave with constant amplitude but random phase.

Solution. From Equation (3.80) we have

$$R(t_1, t_2) = E[X(t_1)X(t_2)] = \int_{-\infty}^{\infty} A\cos(\omega t_1 + \varphi)A\cos(\omega t_2 + \varphi)\frac{1}{2\pi}d\varphi$$

$$= \int_{-\pi}^{\pi} \frac{A^2}{2}[\cos(\omega t_1 - \omega t_2)$$

$$- \cos(\omega t_1 + \omega t_2 + 2\varphi)]\frac{1}{2\pi}d\varphi$$

$$= \int_{-\pi}^{\pi} \frac{A^2}{2}\cos(\omega t_1 - \omega t_2)\frac{1}{2\pi}d\varphi$$

$$- \int_{-\pi}^{\pi} \frac{A^2}{2}\frac{1}{2\pi}\cos(\omega t_1 + \omega t_2 + 2\varphi)d\varphi \qquad (3.85)$$

$$= \frac{A^2}{2}\cos(\omega t_1 - \omega t_2)$$

where the integral over φ has been reduced to run over its range, simple trigonometric relations have been used to go from line one to line two, and the second term on the right-hand side of line three is zero.

Three concepts that are of particular importance in the theory of stochastic processes are now introduced. A stochastic process $X(t)$ is called *strict-sense stationary* if its statistical properties are unchanged with a shift of the origin. It is called *wide-sense stationary* if its mean is constant and its autocorrelation function depends only on $\tau \equiv t_1 - t_2$. In both of these cases we have $E[X^2(t)] = R(0)$. Figure 3.14 illustrates a stationary stochastic process and two types of non-stationary behaviour.

In dealing with stochastic processes it may well be that it is easier to measure a long time series of one realisation of a process than it is to measure a number of different realisations. The question then arises whether the time average over a long period of time of a single realisation is equivalent to averaging over several different realisations. A stochastic process $X(t)$ is called *ergodic* if its ensemble averages equal appropriate time averages.

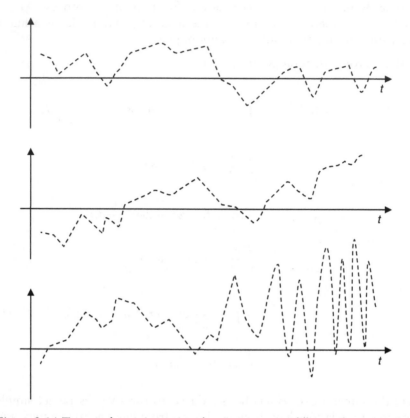

Figure 3.14 Top panel – stationary stochastic process; middle panel – nonstationary stochastic process with a time-dependent mean; bottom panel – nonstationary stochastic process with a time-dependent mean and variance.

The *power spectrum* (or spectral density), $S(\omega)$, of a process $X(t)$ is the Fourier transform of its autocorrelation:

$$S(\omega) = \int_{-\infty}^{\infty} R(\tau)e^{-i\omega\tau}\,d\tau \tag{3.86}$$

and

$$R(\tau) = \frac{1}{2\pi} \int_{-\infty}^{\infty} S(\omega)e^{i\omega\tau}\,d\omega \tag{3.87}$$

Example 3.27. A stochastic process has an autocorrelation function given by $R(\tau) = exp[-(t/T)^2]$, where T is the 'correlation' time. Find the power spectrum of the process.

Solution. To calculate the power spectrum we need to find the Fourier transform of the autocorrelation function:

$$S(\omega) = \int_{-\infty}^{\infty} e^{-\left(\frac{\tau}{T}\right)^2} e^{-i\omega\tau} d\tau$$

$$= \int_{-\infty}^{\infty} e^{-\left\{\left(\frac{\tau}{T}\right)^2 + i\omega\tau\right\}} d\tau$$

$$= \int_{-\infty}^{\infty} e^{-\left\{\left(\frac{\tau}{T}\right) + i\omega\tau\right\}^2 - \left(\frac{\omega T}{2}\right)^2} d\tau$$

$$= e^{-\left(\frac{\omega T}{2}\right)^2} \int_{-\infty}^{\infty} e^{-\left\{\left(\frac{\tau}{T}\right) + i\omega\tau\right\}^2} d\tau$$

$$= e^{-\left(\frac{\omega T}{2}\right)^2} \int_{-\infty}^{\infty} e^{-y^2} T dy$$

$$= T\sqrt{\pi} e^{-\left(\frac{\omega T}{2}\right)^2}$$

where the change of variables $y = (\tau/T) + (i\omega T/2)$ has been used.

Example 3.28. It is left as an exercise for the reader to verify that stochastic processes with autocorrelation functions given by (a) an exponential function $R(\tau) = \exp[-|\tau|/T]$, and (b) a modified exponential function $R(\tau) = (1 + |\tau|/T)\exp[-|\tau|/T]$, where T is the 'correlation' time, have power spectra $S(\omega) = 2T/[(1 + T^2\omega^2)]$ and $S(\omega) = 4T/[(1 + T^2\omega^2)^2]$, respectively. The autocorrelation functions and their corresponding power spectra in Examples 3.27 and 3.28 are shown in Figures 3.15 and 3.16.

Example 3.29. Generation of a normal stochastic process with arbitrary autocorrelation function.

The ideas introduced above can be used to simulate realisations of stochastic processes with normal statistics and arbitrary autocorrelation

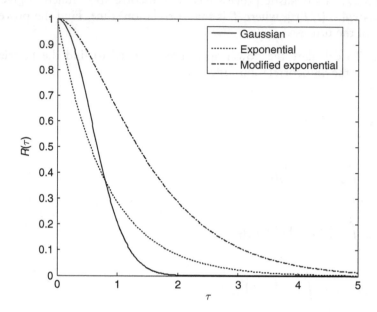

Figure 3.15 Gaussian, exponential and modified exponential autocorrelation functions.

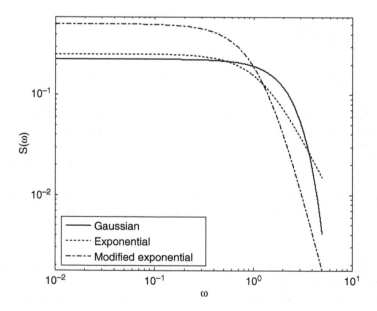

Figure 3.16 Power spectra corresponding to the Gaussian, exponential and modified exponential autocorrelation functions.

function. The idea is outlined below. Implementing this as a computer program is not computationally efficient, however. For details of computational procedures the reader is referred to the publications listed under further reading at the end of this chapter.

Description. Here we consider the simulation of a continuous stochastic variable. Let the stochastic variable have autocorrelation function $R(\tau)$. As this is both real and even, so is its Fourier transform. Thus, we will consider its power spectrum in terms of a Fourier cosine transform:

$$S(\omega) = \frac{2}{\pi} \int_0^\infty R(\tau) \cos(\tau\omega) d\tau \qquad (3.88)$$

and write $A(\omega) = \sqrt{[S(\omega)]}$. Now, choose N equally spaced frequencies $\omega_i = i\Delta\omega$, where N and ω_N are large enough to resolve the features of S sufficiently. Next, choose N independent random phases φ_i, which are uniformly distributed in $(0, 2\pi)$. Now define a function $X(t)$ by

$$X(t) = 2\sqrt{\frac{\Delta\omega}{\pi}} \sum_{i=1}^N A_i \sin(\omega_i t + \varphi_i) \qquad (3.89)$$

where $A_i = A(\omega_i)$. Then $X(t)$ is a continuous function of t with the required statistics, as will be demonstrated below.

The random part of the definition of the function in Equation 3.88 is in the choice of phases. Each different set of phases gives rise to a different realisation of a stochastic process X, and averages can therefore be taken over this ensemble. The ensemble average, denoted by $< . >$, is the integral

$$< . > = \frac{1}{2\pi} \int_0^{2\pi} . \, d\varphi$$

First, it is easy to check that $< X > = 0$. Also, by the central limit theorem, for large N the values of $X(t)$ are normally distributed. It remains to calculate the autocorrelation function of $X(t)$. Let us write $t_i = \omega_i t + \varphi_i$

and $s_i = \omega_i s + \varphi_i$. As φ_i are uniform in $(0, 2\pi)$, using standard trigonometric integrals it is easy to show that:

$$< \sin(t_i) > = 0$$

$$< \sin(t_i) \cos(t_i) > = 0$$

$$< \sin^2(t_i) > = 1/2$$

$$< \sin(t_i) \sin(s_i) > = \frac{\cos(\omega_i \tau)}{2}$$

where $\tau = s - t$. Now, the autocorrelation function can be written as:

$$< X(t)X(s) > = \frac{4\Delta\omega}{\pi} \sum_i^N \sum_j^N A_i A_j < \sin(t_i) \sin(s_j) >$$

$$= \frac{4\Delta\omega}{\pi} \sum_i^N A_i^2 < \sin(t_i) \sin(s_i) >$$

$$= \frac{2\Delta\omega}{\pi} \sum_i^N A_i^2 \cos(\omega_i \tau)$$

$$\cong \frac{2}{\pi} \int_0^\infty S(\omega) \cos(\omega\tau) d\omega$$

$$= R(\tau)$$

as required. Note that in the above derivation it has been assumed that $\sin(t_i)$ and $\sin(s_i)$ are independent.

In many real-life applications it is necessary to relax the requirement of normality. Numerical methods for generating an ensemble of sequences with arbitrary statistics and autocorrelation properties are now becoming available. Some of these are used in the later examples in Chapter 6.

Further reading

Fox, C. G., 1987. An inverse Fourier transform algorithm for generating random signals of a specified spectral form. *Computers and Geosciences*, 13(4): 369–374.

Jenkins, G. M. and Watts, D. G., 1968. *Spectral Analysis and Its Applications*, Holden-Day, San Francisco.

Macaskill, C. and Ewart, T. E., 1984. Computer simulation of two-dimensional random wave propagation. *IMA Journal of Applied Mathematics*, 33: 1–15.

Papoulis, A., 1984. *Probability, Random Variables and Stochastic Processes*, 2nd edition, McGraw-Hill, Electrical Engineering Series, Singapore, p. 576.

4 Extremes

4.1 Introduction to extremes

Rivers, estuaries and coastal regions have always been a popular place for commerce, recreation and habitation. In many countries, land adjacent to waterways is much more valuable than that inland. However, low-lying flood plains are subject to flooding and erosion. Public awareness of the potential for climate change to adversely affect the lives of a huge number of people has been raised in recent years by the news coverage of many high-profile flooding events around the world.

Naturally, there has been a call for the construction of defences to reduce the risk of flooding and erosion. Engineering structures that provide protection against flooding and erosion hazards include open (ditch) and closed (piped) drainage systems, scour protection, embankments, revetments, breakwaters and coastal flood defences. Examples of some of these types of structures are shown in Figure 4.1.

The design of these structures usually takes into account the anticipated lifetime of any human, natural or built assets. Thus, a new hotel being built near an eroding seafront cliff may have a planned life of 50 years. Knowing the typical annual cliff recession rate, one could position the hotel at least 50 times this distance landward of the current cliff position. The economic planning of the investment in the hotel would be designed to recoup all costs well before the 50-year limit. Protection of larger or more diffuse assets, such as towns or highly productive agricultural land, provides a greater challenge. Nevertheless, current planning in the majority of countries uses some measure of the benefit accruing from the construction of the defence relative to the cost of its construction and maintenance in justifying the construction of defences.

The causes of flooding and erosion are driven predominantly by the weather. As this still remains extremely difficult to predict over periods longer than a few days, there is a high degree of uncertainty in the conditions that may be experienced by a structure. As a result, defences are typically designed to withstand conditions of a specified severity (e.g., the storm conditions encountered once every 50 years on average), judged to

Figure 4.1 Examples of flood defence structures (left to right and from top to bottom): (a) Storm water outflow from seawall at Teignmouth, UK. (b) Harbour breakwater at Nice, France. (c) Concrete sea wall with rock scour protection at Sheringham, Norfolk, UK. (d) Harbour breakwater at Sohar, Oman (courtesy of Halcrow Group Ltd). (e) Low-lying rock groynes on a sandy beach backed by vertical blockwork wall in Colwyn Bay, Wales. (f) Concrete mattress revetment on earth embankment at Freiston, Lincolnshire, UK. (g) Short piped sea outfall at Teignmouth, UK. (h) Sea wall with rock toe protection and short groynes at West Bay, UK (courtesy of Dr David Simmonds).

provide an appropriate balance between cost on the one hand and the level of protection on the other.

It is often the case that a structure has to be designed to resist a condition so extreme that no similar condition may be found in available measurements or records. One way to proceed is to fit a probability distribution to the measurements and extrapolate this to find the conditions corresponding to the rarity of the required event, taking due account of the uncertainty in this process. The range of probability distributions that is used for design and the methods for fitting them to the measurements have been the subject of much study, and the following sections provide an introduction and guide to some of the main techniques currently in use.

Almost all the variables encountered in hydraulic and coastal engineering are continuous functions of time (e.g., wind speed, current speed, wave height). However, measurements are taken at fixed intervals, resulting in a discrete set of values over time or time series. Typically, wave and water-level records are available at hourly or 3-hourly intervals, although this can vary according to the instrument and processing adopted. Suppose we have a time series of significant wave heights, such as those plotted in Figure 4.2a.

Each point of the time series can be considered to be an individual event with a duration equal to the interval between successive points. Depending on the sampling rate and the characteristics of the random variable, consecutive measurements may or may not be independent. For many of the statistical manipulations that will be used later, the events are required to be independent. It is important, therefore, to check that the time series are independent. A convenient way to do this is to calculate the autocorrelation function of the time series (see Chapter 3). For the wave record plotted in Figure 4.2a this is shown in Figure 4.2b. If the autocorrelation function drops rapidly from its value of 1 at zero lag, then the assumption that events are independent is reasonable. If the autocorrelation drops more slowly, this shows a similarity or correlation between consecutive values. Correlation of wave records has been investigated and for UK waters, measurements separated by more than 12 hours may be treated as being independent to a good approximation (HR 2000). On the basis of the short record in Figure 4.2a, a value of 12 hours looks a reasonable choice for this location too.

For the case of pluvial and fluvial loading, theoretical and empirical evidence suggest that rainfall patterns exhibit so-called 'multifractal' or 'scaling' behaviour (Lovejoy et al. 1996). This implies that rainfall has features existing on a continuum of spatial scales, from many kilometres down to tens of metres. Zawadzki et al. (1994), Grecu and Krajewski (2000),

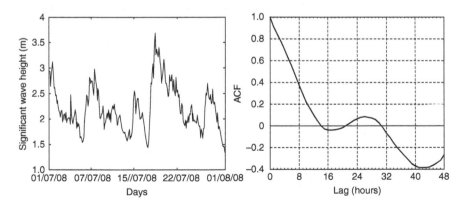

Figure 4.2 (a) Time series of significant wave heights from NOAA wave buoy 42058, situated at 15°5′33″N 75°3′52″W (Central Caribbean – North of Colombia). (b) The autocorrelation function of the wave height record.

Germann and Zawadzki (2002) demonstrated that the lifetime of precipitation features is approximately proportional to their size. Thus, while the smallest features observed by operational weather radar generally persist for no more than a matter of minutes, the largest features may persist for many hours or even days. At catchment scale, a similar general observation to that made for wind waves applies, although it is always advisable, *at the very least,* to plot and look at your data (and preferably analyse it for correlation properties) rather than simply *assume* that a particular correlation property holds.

It is yet to be defined what is meant by *extreme*. One possibility is to define extreme events as those that are greater than some threshold value. The statistics of the values over the threshold may be studied using the *peaks over threshold* method described later. Alternatively, we may decide to 'block' the data into sections of a fixed length, say 1 year, and select the maximum value from each block and investigate the properties of these maxima.

In design, a structure is typically conceived so as to withstand an event of a specified severity – that is, the *design condition*. For events of a severity up to and including the design condition, the structure is expected to provide adequate protection. Conversely, for events of a severity exceeding the design condition, the ability of the defence to provide protection cannot be assumed. To determine the chance of this occurrence we require the probability of the design condition being exceeded. This is most easily obtained from the cumulative distribution; but the cumulative distribution of what? Due to the inherent time scales, a useful 'unit' of time is a year.

Using the cumulative distribution function of a variable, questions relevant to flood or damage prediction may be addressed. For example, in the case of river level it might be required to answer the following questions:

i   What is the probability that the maximum level in a particular year is more than $h \, \mathrm{m^3/s/m}$?
ii   In the next n years what is the probability that the highest level in the n years will be less than $h \, \mathrm{m^3/s/m}$?
iii   What is the 1 in N year annual maximum level?

If $F(h)$ is the distribution function of the annual maxima of river levels, then the answers are: (i) $1 - F(h)$; for (ii), if the annual maxima are independent, then the required probability is Prob{the maximum for each year is less than h} $= F(h)^n$; for (iii) we need the concept of a *return period.*

In engineering design the usual measure of the rarity or severity of an event is the *return period,* T. The *T-year event* is the event that has a $1/T$ chance of being exceeded in any given year. It may be linked to the distribution function as follows. Suppose we have a time series containing N independent values of a load variable $L_1, L_2, L_3, \ldots, L_i, \ldots, L_N$ at intervals of Δt, and we have a loading threshold of L_t, above which the structure cannot perform its intended purpose. Of the N values, assume m are greater than L_t. Approximating the probability of L_i exceeding L_t by $m/N \approx 1 - F(L_t)$. Equivalently, this may be interpreted as meaning there will be m exceedances every N events (or over $N \cdot \Delta t$ units of time) on average, that is, one exceedance every N/m events or $\frac{N \cdot \Delta t}{m}$ units of time. $\frac{N \cdot \Delta t}{m}$ is the *return period* and is usually expressed in units of years. Thus, the answer to (iii) is the value h such that $F(h) = 1/N$, or, $h = F^{-1}(1/N)$.

Example 4.1. A time series of wave heights is sampled at 12-hourly intervals and contains 292,200 independent values. Of the values recorded only four exceed a given threshold wave height, H_t. What is the return period corresponding to the wave height H_t?

Solution. We have $\Delta t = 12$ hours, $N = 292,200$ (corresponding to 400 years), and $m = 4$. The return period is

$$\frac{292200}{4} \times 12 \times \frac{1}{24 \times 365.25} = 100\,\text{years}$$

The last term on the left-hand side converts units of hours to years (24 hours per day and 365.25 days per year).

A concept related to, but distinct from, the return period is the *design life* of a structure. The *design life* of a structure is the period over which the structure is expected to continue providing protection against the design condition. The return period is not the same as the design life. For example, a particular sewer pipe may be designed to last 50 years, but to be able to carry the 1 in 25 year flow. The fact that a storm leads to extreme rainfall in excess of the 1 in 25 year value does not necessarily mean the sewer will be broken, and therefore fail to continue working after the particular storm. Rather, it means that there will be flooding because the sewer cannot convey all the water fast enough. After the flooding subsides, the sewer should continue to provide protection against the 1 in 25 year flow.

Over a period of 50 years the integrity of the pipe is likely to deteriorate. An adequate design will take such deterioration into account, so that at the beginning of its life a structure can probably withstand an event rather more severe than the design condition. Gradual deterioration over time will impair its performance, so that, at the end of its design life, it still provides protection against the original design condition, as illustrated in Figure 4.3.

Considering annual maxima, the probability of exceedance (P), the design life (L) and the return period (T) are related by:

$$P = 1 - \left(1 - \frac{1}{T}\right)^L \tag{4.1}$$

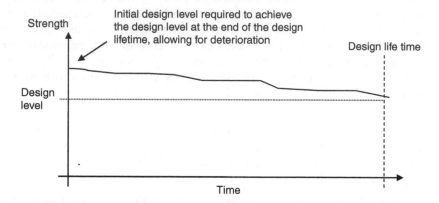

Figure 4.3 Design to resist an extreme event accounting for material deterioration over the design life.

Figure 4.4 plots the return period against duration for fixed values of probability of exceedance.

It can be seen that, if the return period of an extreme event is the same as the design life, there is a ~63% chance that the extreme event will be exceeded during the period of the design life.

In the preceding discussion, the form of the distribution function that might best describe the extreme values of a random process has been avoided. A brief background to the theory of extreme values, together with

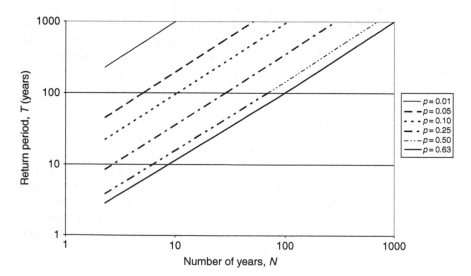

Figure 4.4 Exceedance probability as a function of event return period and duration.

details of fitting these to data and estimating values corresponding to large return periods, are the topics of the remainder of this chapter.

4.2 Limiting distributions and behaviour

We start by considering the extremes of the normal distribution, because it is analytically tractable and illustrates several important features. If the random variable X is normally distributed with mean μ and standard deviation σ then it obeys

$$F(x) = \int_{-\infty}^{x} \frac{1}{\sigma\sqrt{2\pi}} e^{-\left(\frac{(t-\mu)}{\sigma\sqrt{2}}\right)^2} dt$$

In the case of n variables, $X_1, X_2, \ldots, X_n$, which are independent and have the same normal distribution, then (see Chapter 3) the distribution of the maximum of $X_1, X_2, \ldots, X_n$ is

$$F_{X\max}(x) = \left(\int_{-\infty}^{x} \frac{1}{\sigma\sqrt{2\pi}} e^{-\left(\frac{(t-\mu)}{\sigma\sqrt{2}}\right)^2} dt \right)^n$$

Figure 4.5 illustrates how the distribution of the maximum changes with increasing n. Two facets of the distribution are clear. First, it is not normal and, second, the mean of the distribution progressively increases. In addition, it may be shown that, as n increases, this distribution approaches zero for every finite value of X. Some standardisation is therefore necessary. The history behind the theory of extreme value distributions can be traced to

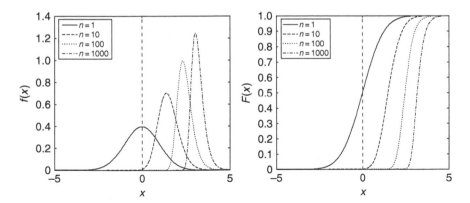

Figure 4.5 Probability density (left) and distribution (right) function of n independent normally distributed variables as a function of *n*.

Fisher and Tippett (1928), who sought distributions for the maxima that satisfied

$$F^n(x) = F(a_n x + b_n) \tag{4.2}$$

where $a_n > 0$ and b_n are arbitrary constants, that is, the distribution of the maxima is the same, up to a linear re-scaling, as F itself. Fréchet (1927) found a solution for the special case of $b_n = 0$. Fisher and Tippett (1928) found three classes of solution to Equation (4.2), which can be written in a single unifying form:

$$F(x; \mu, \sigma, \delta) = \exp\left[-\left\{1 - \delta\left(\frac{x - \mu}{\sigma}\right)\right\}^{1/\delta}\right] \tag{4.3}$$

which has the corresponding probability density function

$$f(x; \mu, \sigma, \delta) = \frac{1}{\sigma}\left\{1 - \delta\left(\frac{x - \mu}{\sigma}\right)\right\}^{(1/\delta) - 1} \exp\left[-\left\{1 - \delta\left(\frac{x - \mu}{\sigma}\right)\right\}^{1/\delta}\right] \tag{4.4}$$

with $\quad 1 - \delta\left(\frac{x - \mu}{\sigma}\right) > 0; \quad -\infty < \mu, \delta < \infty; \quad 0 < \sigma < \infty$

Thus, the data are bounded below if $\delta < 0$ and bounded above if $\delta > 0$. This distribution is known as the *generalised extreme value* (GEV) distribution, and sketches of its shape are shown in Figure 4.6. A rigorous mathematical proof that this is the only nontrivial solution of Equation (4.2) may be found in Gnedenko (1943) and De Haan (1976).

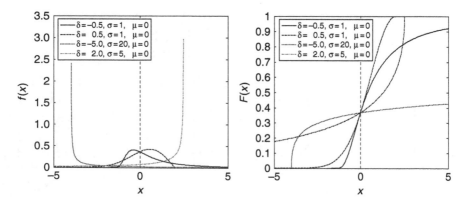

Figure 4.6 GEV probability density function (left) and distribution function (right) for various parameter values.

The three classes of solution correspond to $\delta > 0$ (*EV3* or *Weibull* distribution), $\delta < 0$ (*EV2* or *Fréchet* distribution) and $\delta = 0$ (*EV1* or *Gumbel* distribution). For future reference, some useful properties of these three distributions are given in detail here.

Gumbel distribution

$$F(x; \mu, \sigma) = \exp\left[-\exp\left(-\frac{x - \mu}{\sigma}\right)\right] \tag{4.5}$$

$\sigma > 0$, and $-\infty < x < \infty$.

Mean $= \mu + 0.57721\sigma$

Variance $= \dfrac{\sigma^2 \pi^2}{6}$

$F^{-1}(p) = \mu - \sigma \ln\left[-\ln(p)\right]$

Random numbers with a Gumbel distribution, g_i, may be generated from random numbers with a standard uniform distribution, u_i, by $g_i = \mu - \sigma \ln[-\ln(u_i)]$. The maximum likelihood estimators of μ and σ are given by:

$$\hat{\mu} = -\hat{\sigma} \ln\left[\frac{1}{n} \sum_{i=1}^{n} e^{-\left(\frac{x_i}{\hat{\sigma}}\right)}\right]$$

$$\hat{\sigma} = \frac{1}{n} \sum_{i=1}^{n} x_i - \frac{\sum_{i=1}^{n} x_i e^{-\left(\frac{x_i}{\hat{\sigma}}\right)}}{\sum_{i=1}^{n} e^{-\left(\frac{x_i}{\hat{\sigma}}\right)}} \tag{4.6}$$

Distributions whose extremes lead to a Gumbel distribution include normal, log-normal, exponential and gamma. Plots of the Gumbel density and distribution functions are shown in Figure 4.7.

Fréchet distribution

$$F(x; \mu, \sigma, \gamma) = \exp\left[-\left(\frac{x - \mu}{\sigma}\right)^{\gamma}\right] \qquad x \geq \mu \tag{4.7}$$

$\sigma > 0$ and $\gamma > 0$.

Mean $= \mu + [\sigma \Gamma(1 - 1/\gamma)] \qquad \gamma > 1$

Variance $= \sigma^2[\Gamma(1 - 2/\gamma) - \Gamma^2(1 - 1/\gamma)] \qquad \gamma > 2$

$F^{-1}(p) = \mu + \sigma\left[-\ln(p)\right]^{-1/\gamma}$

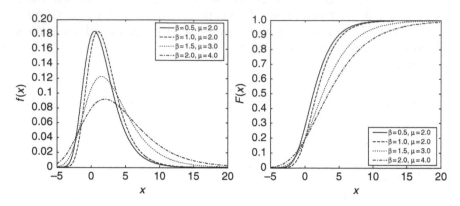

Figure 4.7 Gumbel probability density (left) and distribution (right) functions for various parameter settings.

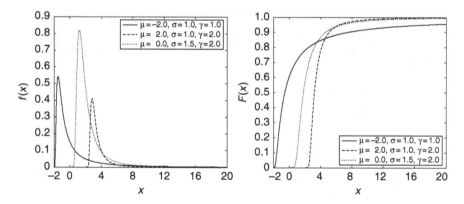

Figure 4.8 Fréchet probability density (left) and distribution (right) functions for various parameter settings.

Distributions whose extremes lead to a Fréchet distribution include the Pareto, log-gamma and Cauchy distributions. Moments of order $k \geq \gamma$ do not exist, which can complicate the estimation of parameters σ and γ. Plots of the Fréchet density and distribution functions are shown in Figure 4.8.

Weibull three-parameter distribution

$$F(x; \mu, \sigma, \alpha) = \exp\left[-\left(\frac{\mu - x}{\mu - \sigma}\right)^\alpha\right] \qquad x \leq \mu \tag{4.8}$$

where μ is the upper bound of X, $\sigma < \mu$ is a location parameter and $\alpha > 0$ is a shape parameter.

$$\text{Mean} = \mu - (\mu - \sigma)\Gamma\left(\frac{1+\alpha}{\alpha}\right)$$

$$\text{Variance} = (\mu - \sigma)^2\left[\Gamma\left(\frac{2+\alpha}{\alpha}\right) - \Gamma^2\left(\frac{1+\alpha}{\alpha}\right)\right]$$

$$F^{-1}(p) = \mu - (\mu - \sigma)\left[-\ln(p)\right]^{1/\alpha}$$

Distributions whose extremes lead to a Weibull distribution include the uniform and beta distributions.

Corresponding distributions for the *minimum* of n independent and identically distributed variables X_i can be derived. The most widely used in hydraulic and reliability analysis is the Weibull distribution for *minima*. It has the following distribution function:

$$F(x; \varepsilon, \sigma, \beta) = 1 - \exp\left[-\left(\frac{x-\varepsilon}{\sigma-\varepsilon}\right)^\beta\right] \qquad x \geq \varepsilon \qquad (4.9)$$

where ε is the lower bound of X, and $\beta > 0$ is a shape parameter. We also have:

$$\text{Mean} = \varepsilon + (\sigma - \varepsilon)\Gamma\left(\frac{1+\beta}{\beta}\right)$$

$$\text{Variance} = (\sigma - \varepsilon)^2\left[\Gamma\left(\frac{2+\beta}{\beta}\right) - \Gamma^2\left(\frac{1+\beta}{\beta}\right)\right]$$

$$F^{-1}(p) = \varepsilon - (\sigma - \varepsilon)\left[-\ln(1-p)\right]^{1/\beta}$$

Estimation of the parameters for the three-parameter Weibull distribution is not straightforward. If the upper or lower bound is known or can be estimated on the basis of other information, then this fixes one of the parameters and the method of moments can be used. For the case when $\varepsilon = 0$, the maximum likelihood method can be used with the estimators of σ and β given by the simultaneous solution of:

$$\hat{\sigma} = \left[\frac{1}{n}\sum_{i=1}^{n} x_i^{\hat{\beta}}\right]^{1/\hat{\beta}}$$

$$\hat{\beta} = \frac{n}{\left(\frac{1}{\hat{\sigma}}\right)^{\hat{\beta}}\sum_{i=1}^{n} x_i^{\hat{\beta}}\ln(x_i) - \sum_{i=1}^{n}\ln(x_i)} \qquad (4.10)$$

Plots of the Weibull density and distribution functions for maxima are shown in Figure 4.9.

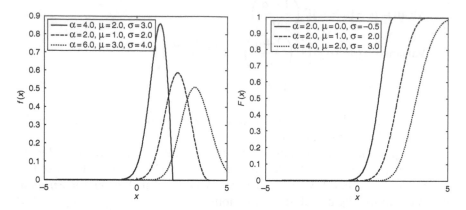

Figure 4.9 Weibull probability density (left) and distribution (right) functions for various parameter settings.

A second method of describing extremes is to set a threshold value, and to define the exceedances of this value as 'extreme'. This approach is often called 'peaks over threshold' (POT) in Europe and 'partial duration series' (PDS) in the USA. Variables X_i with a distribution function F are considered extreme events if a threshold u is exceeded. Using essentially the same arguments that justified the GEV asymptotic form for maximum values, it may be shown that the natural family of distributions to describe the exceedances of a threshold is the Generalised Pareto Distribution (Pickands 1975). Thus, the conditional distribution of the random variable X, given that $X > u$, is

$$\Pr(X \leq x \mid X > u) = F(X \leq x / X > u), \qquad x > u \qquad (4.11)$$

where F is the distribution function of X. For u sufficiently large this can be well approximated by the Generalised Pareto Distribution, $\mathrm{GPD}(\sigma, \xi)$,

$$F(X \leq x \mid X > u) = 1 - \{1 - \xi\,(x - u)\,/\sigma\}^{1/\xi}, \qquad x > u \qquad (4.12)$$

where σ is a scale parameter ($\sigma > 0$) and ξ is a shape parameter. If $\xi = 0$, we retrieve the exponential distribution. The GPD pdf is shown in Figure 4.10 for several values of the parameters.

The POT/GPD approach is widely considered as being much less wasteful of information in comparison with the block maximum/GEV method. Nevertheless, it requires a judicious selection of the threshold value, some check that for a given threshold the peak values are independent, and a means of converting the result into a form suitable for engineering design and assessment (i.e., a value corresponding to a particular return period). The choice of threshold is not straightforward and a survey of methods available

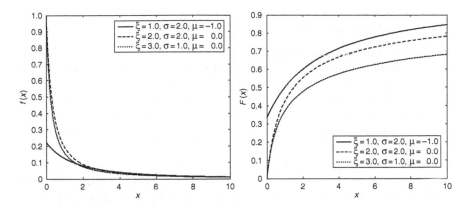

Figure 4.10 GPD probability density (left) and distribution (right) functions for various parameter settings.

for addressing the first two of these points may be found in Thompson (2009). For the last point, the following 'conversion' relations are required:

$$
X_N = \begin{cases} u + \dfrac{\sigma}{\xi} \left[(Nn_y p)^\xi - 1 \right] & \xi \neq 0 \\[2ex] u + \ln(Nn_y p) & \xi = 0 \end{cases}
\tag{4.13}
$$

where X_N is the N-year return level, n_y is the average number of excesses per year, and p is the probability that $X > u$. It should not be surprising that the GEV and GPD distributions have some similarities. Indeed, Davison and Smith (1990) showed that, if the exceedances obey a GPD(σ, ξ) distribution, then the distribution of the annual maxima of the observations is described by a GEV distribution with parameters defined in terms of σ, ξ, n_y and p. Thus, knowing the GPD parameters, we may estimate the corresponding GEV parameters using all the large values rather than just the annual maximum values.

In practice, if the standard errors associated with fitting annual maxima to a GEV distribution are sufficiently small to allow specification of design conditions within acceptable bounds, then this method is often used. It is also worth noting, on a practical level, that some historical recordings (e.g., tide gauge records) only archive the annual maximum values. In this case the GEV distribution is the sensible option. However, with the introduction of continuous digital recording, continuous time series records are becoming the norm and the POT method is a viable alternative.

4.3   Fitting extreme distributions to data

4.3.1   *Least-squares fitting*

In Chapter 3 the concept of least squares was introduced as a means of estimating the parameters of a distribution that best fits the observed data in
the sense of minimising the mean square differences in the y-direction. In
Examples 3.24 and 3.25 it was shown how the method for fitting a straight
line to observations could be adapted to more complicated functions by
transformations of the independent and dependent variables. Expressions
for estimating the parameters of the Gumbel and two-parameter Weibull
distributions were given in the previous section. In this section we return to
the process of fitting a curve (in particular a probability distribution function) to observations. The discussion starts with the normal distribution and
then progresses to the choice of plotting position and estimation of standard
errors.

All cumulative distribution functions are nondecreasing functions of the
random variable. The idea behind probability plotting is to transform the
vertical (and sometimes the horizontal) scale so that the distribution function plots as a straight line. For a normal distribution it is necessary to alter
only the vertical scale. Consider a normal random variable X with a mean μ
and a standard deviation σ. With the wide availability of computer spreadsheet programs, it is easy to recreate what used to be done on commercially
produced 'probability paper' by performing the mathematical transformation and plotting the results. Recall that the standard normal form Z of a
normal random variable X is

$$Z = \frac{X - \mu}{\sigma} = \left(\frac{1}{\sigma}\right) X + \left(\frac{-\mu}{\sigma}\right)$$

For any realisation x of X, the standardised value, z, is defined in the
corresponding way. Thus,

$$F_X(x) = p = \Phi\left(\frac{x - \mu}{\sigma}\right) \tag{4.14}$$

Taking the inverse of Equation (4.14) yields

$$\Phi^{-1}(p) = z = \left(\frac{1}{\sigma}\right) x + \left(\frac{-\mu}{\sigma}\right) \tag{4.15}$$

Now, Equation (4.15) represents a linear relationship between z and x.
The slope of the line is the reciprocal of the standard deviation, and the
intercept is minus the mean divided by the standard deviation. Note that, in
order to make the graph easier to use, the y-axis is often shown as probabilities (p), and stretched, rather than plotting the z values on a linear

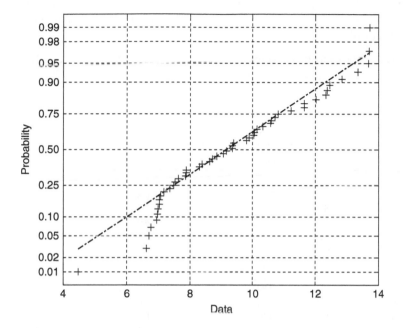

Figure 4.11 Straight-line plot on normal probability 'paper', demonstrating that on the scaled ordinate the normal curve becomes a straight line.

scale. This relationship is illustrated in Figure 4.11. This format is easier for determining quantiles. For example, the values of x corresponding to $F_X(x) = 0.25$, 0.5 and 0.75 are the 25% quartile, the median and the 75% quartile, respectively.

The application of this to a set of observations is as follows:

1 Sort the observations in increasing order, with $x_1 \geq x_2 \geq \ldots \geq x_i \geq \ldots \geq x_N$, and retain any repeated values.
2 Associate with each x_i a cumulative probability p_i defined as:

$$p_i = \frac{i}{N+1} \tag{4.16}$$

3 For each p_i calculate $z_i = \Phi^{-1}(p_i)$.
4 Plot the coordinates (x_i, z_i) on standard linear axes.
5 If the x_i follow a normal distribution then the coordinates will follow a straight line. The best-fit straight line yields the parameters (mean and standard deviation) of the distribution as described above. If the points do not fall on a straight line, then a normal distribution is probably not appropriate.

Example 4.2. Consider the following set of 12 observations: {5.6, 4.8, 7.6, 12.3, 8.3, 6.4, 5.8, 7.9, 5.1, 6.6, 6.9, 5.9}. Plot the observations on a normal probability plot.

Solution. We follow the steps (1) to (3) above, noting that $N = 12$, and summarise the results in Table 4.1 below.

Table 4.1 Data for Example 4.2

Index	x_i	$p_i = i/(N+1)$	$z_i = \Phi^{-1}(p_i)$
1	4.8	0.08	−1.41
2	5.1	0.15	−1.03
3	5.6	0.23	−0.74
4	5.8	0.31	−0.49
5	5.9	0.38	−0.31
6	6.4	0.46	−0.10
7	6.6	0.54	0.10
8	6.9	0.62	0.31
9	7.6	0.69	0.49
10	7.9	0.77	0.74
11	8.3	0.85	1.03
12	12.3	0.92	1.41

Figure 4.12 shows the points and a best fit-line through them on normal probability axes. The slope is 0.3773, so the standard deviation is $1/0.3773 = 2.65$; the intercept is −2.616, so the mean is $2.616 \times 2.65 = 6.93$. By direct calculation, the mean of the values in Table 4.1 may also be found to be 6.93.

In this example it is clear, even from a visual inspection, that the fit is not very good. Indeed, if the largest value were omitted, a much better fit would have been achieved. The existence of unusual points, or 'outliers', is often an indication of the need for further investigation, either of the quality of the data or a better understanding of the processes being modelled, or of the (in)appropriateness of the choice of probability distribution selected to fit the data.

The method of least squares is one means of performing step (5), that is, to define the line that best fits the observations. The plotting formula Equation (4.16) was proposed by Gumbel (1954). Other formulae have been proposed, for example

$$\frac{i-1}{N}, \quad \frac{i-0.3}{N+0.4}, \quad \frac{i-0.44}{N+0.12}, \quad \frac{i-0.5}{N} \tag{4.17}$$

The last is known as the 'Hazen formula' and has the property of generally being the least biased for larger samples with $N > 20$. The second

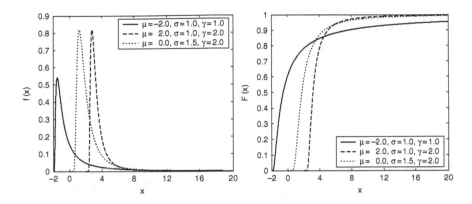

Figure 4.12 Normal probability plot for Example 4.2 plotted on linear axes.

formula is generally favoured for use with smaller samples. Further discussions can be found in Cunnane (1978) and Chambers et al. (1983). The choice of plotting position is important, particularly if it is expected that the observations are *censored*. An observation is said to be *censored* if its exact value is not known, but rather that it lies in some interval. Two common cases are left- and right-censoring. Left-censoring, where it is known that an observation lies in an interval $(-\infty, a)$ for a given constant a, may occur if there is simply no sufficiently extreme event recorded. Right-censoring can occur if the recording equipment is limited in range or breaks down at the extremes of the observations; the observation is in the breakdown range but its precise value is not known. In any case, if there are suspected outliers, it can be wise to treat the extreme observations as right-censored rather than observed to allow for the possibility of equipment breakdown. The relevance of this to plotting position is that, in the case of right-censored data, the plotting position may be replaced by the Kaplan–Meier estimator:

$$p_i = 1 - \frac{N+0.5}{N} \prod_{\substack{j \in I \\ j \leq i}} \frac{N-j+0.5}{N-j+1.5} \tag{4.18}$$

as recommended by Chambers et al. (1983).

4.3.2   *Method of moments*

For cases where only uncensored data are available, the simple method of moments provides a straight forward means of estimating parameters. The idea behind this approach is to equate the sample and population moments, and was illustrated in Example 3.23 for the Rayleigh and Gamma distributions. The method is most effective for distributions that have closed-form expressions for their moments. Here we give an example using the Gumbel, Pareto and two-parameter Weibull distributions.

Example 4.3. The data in Table 4.2 are the monthly maximum wave heights from a wave buoy located near Alghero for the years 1990–2005. Find the best-fit parameters for the data for the Gumbel, Pareto and Weibull distributions, and thus determine the 50-year return wave height.

Solution.   We extract the annual maxima from the dataset {7.5, 8.9, 7.8, 9.1, 9.2, 8.0, 7.6, 9.2, 9.88, 8.11, 8.3, 6.5, 8.2, 5.32}. The mean and standard deviation of the sample annual maxima data are 8.101 and 1.144, respectively. The mean and standard deviation of the Gumbel distribution are:

$$\text{Mean} = \mu + 0.57721\sigma$$

$$\text{Variance} = \frac{\sigma^2 \pi^2}{6}$$

Equating the sample and population moments allows us to solve for σ directly ($\sigma = 0.892$) and then substitute this value into the expression for the mean to obtain $\mu = 7.586$.

The distribution function, mean and standard deviation of the Pareto distribution are:

$$F(x; a, c) = 1 - \left(\frac{a}{x}\right)^c \qquad a \leq x < \infty$$

$$\text{Mean} = \frac{ca}{(c-1)}, \, c > 1 \tag{4.19}$$

$$\text{Variance} = \frac{ca^2}{(c-1)^2(c-2)}, \, c > 2$$

Using the expressions for the population moments, we can eliminate a to obtain a quadratic equation for c. This has the roots -6.153 and 8.153. We select the larger one, as this gives $c > 1$ as required. Using this value in the expression for the mean allows us to solve for a, yielding $a = 7.107$.

We use the form of Weibull two-parameter distribution associated with minima (set $\varepsilon = 0$ in Equation 4.9), because there is a known minimum

Table 4.2 Monthly maximum wave heights from January 1990 to December 2005, recorded at Alghero, Italy

Month	Height	Month	Height	Month	Height	Month	Height	Month	Height
1	3.4	37	5.6	73	5.5	109	7.55	145	5
2	5.7	38	6.7	74	7.9	110	7.9	146	5.52
3	5.2	39	4.9	75	4.5	111	5.48	147	3.84
4	5.5	40	8.6	76	5.8	112	5.81	148	3.23
5	1.8	41	1.6	77	5.5	113	4.4	149	3
6	3.4	42	3.6	78	4.7	114	5.07	150	3.01
7	3.9	43	4.2	79	5.8	115	4.17	151	4
8	2.9	44	3.2	80	2.4	116	2.16	152	5.2
9	3.4	45	3.6	81	6.6	117	3.14	153	4.6
10	3.5	46	6	82	5.2	118	4.39	154	4.8
11	6.2	47	7.6	83	7.3	119	7.84	155	6.5
12	7.5	48	9.1	84	6.6	120	9.88	156	6.4
13	4.1	49	9.2	85	5	121	6.08	157	7.1
14	4	50	4.6	86	6.7	122	6.53	158	8.2
15	3.3	51	4.7	87	5.4	123	8.11	159	4.22
16	7.4	52	8.4	88	5.1	124	3.76	160	4.74
17	5.4	53	3.8	89	6.2	125	2.21	161	4.37
18	4.1	54	4.3	90	2.8	126	3.61	162	2.66
19	3.3	55	2.8	91	3.9	127	4.24	163	2.74
20	2.7	56	3.6	92	4	128	2.82	164	2.42
21	3.4	57	3.8	93	3.2	129	4.13	165	4.08
22	5	58	2.9	94	6.4	130	5.83	166	5.23
23	7.1	59	5.6	95	5.8	131	5.32	167	4.28
24	8.9	60	5.4	96	7.6	132	7.1	168	4.84
25	4.2	61	8	97	9.2	133	7.93	169	4.82
26	6.4	62	6.4	98	3.6	134	8.3	170	4.94
27	6.3	63	6.9	99	6	135	5.84	171	5.32
28	4.2	64	6.1	100	5.2	136	5.79	172	4.23
29	5.7	65	7.4	101	3.5	137	3.07	173	4.55
30	2.7	66	4.6	102	6.9	138	4.7	174	3.99
31	2.3	67	2.1	103	6	139	4.6	175	2.97
32	2.3	68	5.7	104	4	140	3.43	176	3.14
33	3.6	69	4.4	105	6.4	141	4.89	177	4.76
34	3.9	70	1.4	106	5.2	142	2.58	178	3.35
35	6.4	71	6.8	107	5.7	143	5.24	179	4.63
36	7.8	72	6.1	108	6.3	144	7.54	180	5.08

threshold of 0 that all annual maximum wave heights *must* exceed. This gives,

$$\text{Mean} = \sigma \Gamma \left(\frac{1+\beta}{\beta} \right)$$

$$\text{Variance} = \sigma^2 \left[\Gamma \left(\frac{2+\beta}{\beta} \right) - \Gamma^2 \left(\frac{1+\beta}{\beta} \right) \right]$$

No closed-form solution for these simultaneous equations is available, so they have to be solved numerically using a Newton–Raphson or minimisation technique (see e.g., Press et al. 2007). We find $\sigma = 8.580$ and $\beta = 8.435$.

Figure 4.13(a–c) shows the observations and the best-fit distribution on Q-Q plots for the Gumbel, Pareto and Weibull distributions. Note that the scales for each plot are different, as the coordinates have been transformed to yield a straight line for each distribution, as described in Table 4.3.

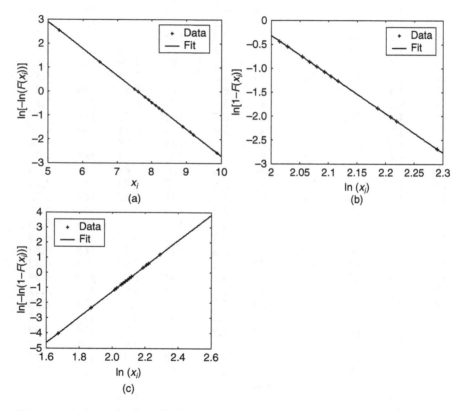

Figure 4.13 Data plus best-fit distributions for Example 4.3: (a) Gumbel; (b) Pareto and (c) Weibull.

Table 4.3 Scaled axes used for the Q-Q plots for the three distribution functions

Model	x	y
Gumbel	x	$\ln[-\ln(F(x))]$
Pareto	$\ln(x)$	$\ln[1 - F(x)]$
Weibull	$\ln(x)$	$\ln[-\ln(1 - F(x))]$

The 50-year return wave height is obtained by setting the distribution function to $1 - 1/50 = 1 - 0.02 = 0.98$ and solving for 'x'. Thus, for the Gumbel distribution:

$$x = F^{-1}(p) = \mu - \sigma \ln\left[-\ln(0.98)\right] = 11.1\,\text{m}$$

Similarly, for the Pareto and Weibull distributions the 50-year wave heights are $11.5\,\text{m}$ and $10.1\,\text{m}$, respectively. The spread in the estimates should not be surprising, because the fit to the observations is good but not perfect in any of the cases, as can be seen from the Q-Q plots. The decision of which of the three statistical distribution models is 'best' is dependent upon, amongst various factors, the standard errors in the estimates of the extreme values and the goodness-of-fit of each distribution to portions of the data. For example, the Gumbel distribution might provide the best fit to the data for points at the lower end of the range, while the Weibull distribution fits the points at the upper end of the range better. If interest is in extremes at the upper end, there could be an argument to prefer the estimates of extreme levels based on the Weibull distribution in this case. Similarly, the standard errors associated with the estimates made with the Pareto distribution might be much smaller than with the other two distributions. In that case, it could be argued that the estimates from the Pareto distribution should be used, as there is less uncertainty associated with them. One method of determining the standard errors is covered in the next section. The problem of how to determine the goodness-of-fit of a particular distribution to different parts of a dataset is covered in Section 4.3.5.

4.3.3 *Maximum likelihood estimation*

The estimation of parameters using the maximum likelihood estimation was described in Section 3.7.4. Here it can also provide a measure of the errors in these estimates as confidence limits. In general, an estimate of a parameter θ' is given by some function, f, of a random sample from the population, $\mathbf{X} = \{X_1, X_2, \ldots, X_n\}$. For large samples, many estimates may be approximated as normally distributed, and confidence bounds are defined in the form

$$E[f(\mathbf{X})] \pm z_c\sqrt{\text{Var}[f(\mathbf{X})]} \tag{4.20}$$

where z_c is a percentage point from the standard normal distribution reflecting the degree of belief in the limits. For example, the 90% confidence limits are based on the 5% and 95% points of the standard normal distribution [i.e., $z_c = 1.96$, as $\Phi(z_c) = 0.05$]. The problem remaining is how to determine $\text{Var}[f(X)]$. This can be estimated from the observed information matrix. If there are p parameters to be estimated the information matrix is defined as the $p \times p$ matrix with elements $-\partial^2 \ln(L)/\partial\theta_i\partial\theta_j$ evaluated at the best-fit values of the parameters. The inverse of this matrix is the estimated covariance matrix, and the elements on the diagonal are the variances of the parameters. Confidence intervals for the θ_i are given by: $\theta_i' \pm z_c \sqrt{[\mathrm{Var}(\theta_i')]}$, where z_c is the chosen percentage point from the standard normal distribution. In all but the simplest cases, the information matrix has to be evaluated and inverted numerically.

Example 4.4. Using the results of Example 4.3, determine the information matrix, correlation matrix and 90% confidence limits of the best-fit parameters of the Gumbel distribution.

Solution. First, we construct the likelihood function, $L(X_i; \mu, \sigma)$. This is given by

$$L(X_i; \mu, \sigma) = \prod_{i=1}^{N} f_{Gumbel}(X_i; \mu, \sigma)$$

where the X_i, $i = 1, 2, \ldots, N$, are the observed annual maxima, and μ, σ are the best-fit parameters of the Gumbel distribution. We now calculate the log-likelihood function and its partial derivatives:

$$\ln[L(X_i; \mu, \sigma)] = \ln\left\{ \frac{1}{\sigma^N} \exp\left(-\sum_{i=1}^{N} \left(\left[\frac{X_i - \mu}{\sigma} \right] + e^{-\left(\frac{X_i - \mu}{\sigma} \right)} \right) \right) \right\}$$

$$\frac{\partial^2 \ln(L)}{\partial \mu^2} = -\frac{1}{\sigma^2} \sum_{i=1}^{N} e^{-\left(\frac{X_i - \mu}{\sigma} \right)}$$

$$\frac{\partial^2 \ln(L)}{\partial \mu \partial \sigma} = \frac{\partial^2 \ln(L)}{\partial \sigma \partial \mu} = -\frac{N}{\sigma^2} - \sum_{i=1}^{N} \frac{1}{\sigma^2} e^{-\left(\frac{X_i - \mu}{\sigma} \right)} \left\{ \frac{X_i - \mu}{\sigma} - 1 \right\}$$

$$\frac{\partial^2 \ln(L)}{\partial \sigma^2} = \frac{N}{\sigma^2} + \sum_{i=1}^{N} 2 \left(\frac{X_i - \mu}{\sigma^3} \right) \left(e^{-\left(\frac{X_i - \mu}{\sigma} \right)} - 1 \right) - \left(\frac{X_i - \mu}{\sigma^2} \right)^2 e^{-\left(\frac{X_i - \mu}{\sigma} \right)}$$

Now construct the information matrix by evaluating the derivatives at the best-fit values of the parameters to give $J = \left(\begin{smallmatrix} 28.39 & -51.30 \\ -51.30 & 200.1 \end{smallmatrix} \right)$. This is a 2×2 matrix, which can be inverted straightaway to find $\mathrm{Corr} = \left(\begin{smallmatrix} 0.0656 & 0.0168 \\ 0.0168 & 0.0093 \end{smallmatrix} \right)$. Taking $z_c = 1.96$, which corresponds to the 5% and 95% points of the standard normal distribution, yields the following 90% confidence limits on the two Gumbel parameters: $\mu = 7.59 \pm 1.96 \, (0.0656) = 7.59 \pm 0.50$, and $\sigma = 0.892 \pm 1.96 \, (0.0093) = 0.892 \pm 0.189$. Using these values to find the 90% confidence limit for the 50-year return value of the water level estimated with the Gumbel distribution gives 11.1 m as the best estimate, as before, with upper and lower limits of 12.3 m and 9.8 m, respectively. These values comfortably embrace the estimates obtained using the other two distributions. More importantly, the magnitude of the confidence interval is large and indicates a considerable degree of uncertainty. This should not come as a surprise, because only 15 years of measurements have been used to estimate the 50-year level!

4.3.4   Getting more from your data

One of the criticisms of using annual maxima to estimate extreme levels is that it is wasteful of data, particularly if you have good time series available. This is one reason why using peaks over threshold is often preferred where appropriate data are available, as is often the case for data from river gauging stations, rainfall records and so on. However, at coastal locations, particularly in harbours away from nationally controlled tide gauges, it was long the custom to record only the highest water level in any year. Thus, many of the longest records comprise only annual maxima, and so a peak over threshold approach is inappropriate in these cases.

Where more data are available, the annual maxima/GEV approach can be modified to include the r-largest annual values. Smith (1986) developed the theory for a family of statistical distributions for extreme values based on $r > 1$ of the largest annual events and applied it to water level observations in Venice. He found that more of the observations could be used subject to the condition that the value of 'r' had to be chosen carefully to ensure the events were 'sufficiently' independent. Specifically, the GEV distribution is an appropriate model for the distribution of block maxima for stationary dependent sequences, provided there is only short-range dependence (see Leadbetter 1983 or O'Brien 1987).

4.3.5   Model selection

As described in previous sections, the conventional approach to the analysis of extreme values proceeds by fitting the rank-ordered observational data to a cumulative distribution function. There are several methods of fitting a model distribution function to a sample of extreme data, such as a graphical fitting method, least-squares estimation, L-moments estimation (Hosking 1990) and maximum likelihood estimation. The L-moments technique has found application in hydrological studies (e.g., Kjeldsen & Jones 2006). The maximum likelihood estimation has been favoured recently in coastal engineering, partly because it is the only method, apart from numerical Bayesian techniques, that can really be considered a general method in the sense that it is applicable to a wide range of problems. The fitting procedure yields a set of optimised parameters, $\bar{\theta}$, for the distribution function.

After fitting a model to the data with maximum likelihood estimation, it is then necessary to evaluate how well the candidate distribution function describes the sample data. As noted in Section 4.3.3, the information matrix may be used to estimate standard errors, and thus confidence intervals, of the parameters. An alternative is to use re-sampling techniques (Section 3.7) to estimate confidence intervals.

As it happens, re-sampling techniques provide a means of comparing how well different statistical models (i.e., distribution functions) fit the data. Almost as a by-product, estimates for the confidence limits of parameter values and quantiles may be retrieved. First, we define an error norm that will be used to calculate how well a distribution fits the observations. By denoting a single sample by $X = \{x_1, x_2 \cdots x_n\}$ an error norm, $\Delta(\bar{\theta})$, can be defined as the maximum deviation between the empirical distribution and an approximating model $F_{\bar{\theta}}$ with optimised parameters $\bar{\theta}$. Linhart and Zucchini (1986) suggested the following error norm:

$$\Delta(\bar{\theta}) = \max_{1 \leq i \leq n} \left| \left[\frac{i}{(n+1)} \right]^b - F_{\bar{\theta}}(r_i)^b \right| \qquad (4.21)$$

where $r_1, r_2, \cdots, r_n$ are the ordered data X in increasing order. The parameter h controls the emphasis of the fitting algorithm to a particular segment of the data: when $h = 1.0$ the test corresponds to the Kolmogorov–Smirnov test, which places equal emphasis across the whole of the rank-ordered distribution; for $h < 1$, emphasis is placed on the lower tail; and for $h > 1$, emphasis is placed on the upper tail.

Now, by considering a series of B replicates, we can compute the expected value of the error norm (Equation 4.11), which is an improved estimator of the standard error, can be computed:

$$\langle \Delta_i(\bar{\theta}_i) \rangle = \left\langle \max_{1 \leq i \leq n} \left| \left[\frac{i}{(n+1)} \right]^b - G_{\bar{\theta}_i}(r_{ij})^b \right| \right\rangle \qquad (4.22)$$

where $j = 1, 2, \cdots, B$ and $\langle \cdot \rangle$ is the expectation operator.

By generating bootstrap replicates from the original sample, then fitting each replicate to different distribution functions and calculating the expected error norm for a range of values of h, it is possible to gain a detailed picture of which distributions best fit which portions of the data.

Example 4.5. Annual maximum wave heights recorded at the US Army Corps coastal research station at Duck, North Carolina for 1981–2007 are plotted in Figure 4.14.

The bootstrap method was employed to create 500 replications. The error norm in Equation (4.21) was used to calculate the expected error norm for various values of h and are tabulated in Table 4.4. For each bootstrap replicate the distributions were fitted to the data using the maximum likelihood

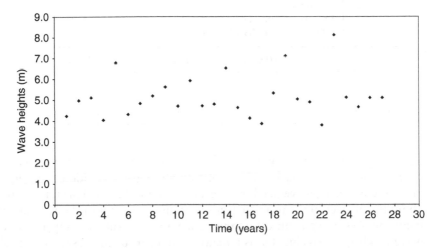

Figure 4.14 Annual maximum wave heights at the US Army Corps of Engineers Coastal Research Station, Duck, North Carolina, from 1981 to 2007 inclusive.

Table 4.4 Expected error norms for various values of h and for four different families of distributions

	$h = 1.5$	$h = 1.25$	$h = 1.00$	$h = 0.75$	$h = 0.50$	$h = 0.25$
Weibull (II)	0.254	0.238	0.214	0.189	0.191	0.177
Weibull (III)	0.177	0.172	0.165	0.158	0.170	0.206
Gumbel	0.167	0.155	0.143	0.131	0.121	0.104
GEV	0.153	0.143	0.134	0.128	0.125	0.117

estimation method and the error calculated accordingly. The errors for fixed distribution and h are then averaged over all replications.

The best fit is obtained with the smallest value of the error norm, and is printed in bold type for each value of h. From the table above it may be seen that the GEV is the best fit, except for the lower end of the data where the Gumbel distribution provides a very slightly better fit. At first glance it might appear puzzling that the Gumbel distribution can perform better than the GEV because the GEV distribution includes the Gumbel distribution. However, it should be noted that the maximum likelihood estimation for three parameters is a much more numerically complex procedure; it is not always easy to find a global maximum and there is greater scope for the accumulation of rounding errors in the calculations. The three-parameter Weibull distribution generally provides a better fit than the two-parameter Weibull distribution. This can be understood because the extra parameter provides an additional degree of freedom with which to fit the observations. Note

Table 4.5 Extreme wave heights at Duck, with 95% confidence limits, for return periods of 10, 50 and 100 years for the best-fit GEV distribution

	10 years	*50 years*	*100 years*
Return value (m)	6.40	7.93	8.65
Lower limit (m)	5.61	6.20	6.46
Upper limit (m)	7.27	10.47	12.62

that we have used the Weibull distribution for minima. This is because the distribution for minima has a 'cut-off' on the left-hand side of the distribution. Obviously, wave heights have a minimum of zero. This information about the physics of the problem leads us to use this form of the Weibull distribution. Assuming a cut-off on the right-hand side (i.e., an upper limit on wave heights) is possible, but is potentially dangerous for design as it could lead to underestimation of extreme wave heights. The Gumbel distribution has only two parameters but still provides a reasonably good fit. In fact, the differences in error norm between the GEV, Gumbel and three-parameter Weibull distributions is actually very small in this case.

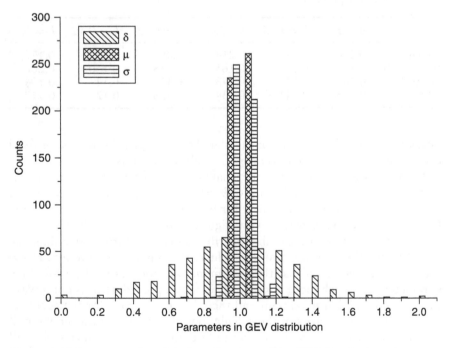

Figure 4.15 Frequency plot of the best-fit values of the GEV distribution parameters obtained from 500 bootstrap replications.

The best-fit distribution can be used to estimate the extreme quantiles corresponding to particular return periods. Table 4.5 summarises the extreme wave heights for return periods of 10, 50 and 100 years. Also included are the 95% confidence limits, determined from the ordered bootstrap replications.

Note the rapid and asymmetric growth in confidence interval with return period, which is indicative of the increasing uncertainty associated with extrapolating the distribution beyond the period covered by the observations.

The bootstrap process also provides 500 sets of parameter values for each of the distributions. These can be used to create a density function for each parameter (see e.g., Li et al. 2008). Figure 4.15 shows a density function for the GEV parameters.

This demonstrates that parameters μ and σ have a small spread in values, while δ has a large spread of values, corresponding to greater uncertainty in the estimated value of this parameter.

Example 4.6. Using the 15 years of monthly maximum wave heights from Alghero (see Figure 4.16), determine the 50- and 100-year return period wave height.

Solution. In this case we have only 15 data points if we select the annual maxima. Alternatively, we could make better use of the data by using peaks over threshold or the r-largest annual events. On the other hand, we could use block maxima – the monthly maxima – to determine the distribution for these, and then scale this accordingly to obtain the annual return periods (i.e., the 10-year return period is the 120-month return period, and so on). Here, we use the latter method. First we calculate the autocorrelation at a lag or 1 month to check whether the data are independent. The lag-1 correlation is 0.36. This is on the high side for strict independence, but for the purposes of this example we proceed on the basis that block maxima and extreme-value analysis are still valid for locally dependent sequences.

The bootstrap method was used to calculate the error norm for various distribution functions. The results are summarised in Table 4.6. This shows that the best fit at the more extreme values is actually found with the Weibull three-parameter distribution (see Figure 4.17).

The values of the Weibull three-parameter distribution corresponding to the best fit are $\alpha = 2.4016$, $\sigma = 4.4496$ and $\mu = 1.0994$. Substituting these values in Equation (4.8), one then sets the value of the distribution function $= 1/(100 \times 12)$ and solves for the value of the wave height. This gives the 100-year wave height as 11.16 m. The bootstrap samples give the 95% confidence limits of this estimate as 9.51 m and 11.87 m, respectively.

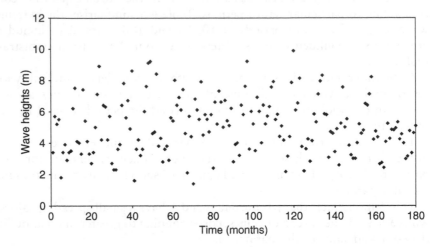

Figure 4.16 Plot of monthly significant wave height maxima at Alghero. Month 0 corresponds to January 1990.

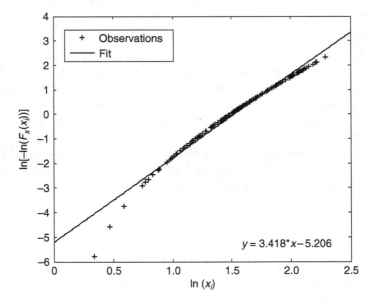

Figure 4.17 Best-fit Weibull distribution together with the data points (*Q-Q* plot).

4.4   Joint extremes

To this point, extremes of a single variable have been discussed. However, in assessing the reliability of a structure or the risk of an undesirable event, one is usually dealing with a problem that involves more than one random variable. For example, the amount of water falling during a storm will

Table 4.6 Bootstrap error norms for Alghero, with 500 bootstrap replicates (see Example 4.6)

	$h = 1.5$	$h = 1.25$	$h = 1.00$	$h = 0.75$	$h = 0.50$	$h = 0.25$
Weibull (II)	0.0689	0.0654	0.0619	0.0615	0.0803	0.1005
Weibull (III)	0.0588	0.0540	0.0499	0.0484	0.0568	0.0877
Gumbel	0.0698	0.0663	0.0620	0.0571	0.0629	0.0886
GEV	0.0622	0.0569	0.0535	0.0529	0.0543	0.0618

depend on the intensity *and* the duration of the storm. Wave overtopping of sea defences is more complicated still, depending upon wave height, wave period, wave direction, water level, beach elevation and the shape and materials of the defence. A full treatment of such problems requires consideration of multivariate extreme values, which is concerned with the joint distribution of extreme values in more than one variable. Such a theory has a wider applicability, because it could be used to examine the dependence between flooding at two neighbouring sites, between the failure of two elements in a network (e.g., a flood embankment composed of elements with different types of construction), or even in the analysis of dependent time series.

The treatment of multivariate extremes is far from trivial. Early progress was made by, amongst others, De Haan and Resnick (1977), Pickands (1981) and Resnick (1987), who proposed a form of theory in which the univariate marginal distributions were the classical extreme-value distributions. The typical line of argument runs as follows. First, assume a model distribution for the marginal distributions of each variable. Second, the choice of marginal distribution defines the form of multivariate distribution, up to an arbitrary function, usually termed the 'dependence function'. Third, this arbitrary function is specified on the basis of an analysis of the dependence properties (usually pairwise) between each variable. Tawn (1988) proposed a number of models for the dependence function. The above process may sound straightforward, but there are many mathematical constraints on the form of the marginal and joint distributions and on the dependence function that need to be observed. Nevertheless, the concepts of block maxima and peaks over thresholds discussed for univariate cases can be carried over to the multivariate case.

More recent development of the theory has seen progress in the application of the peaks over thresholds approach to the bivariate case in particular. By taking the marginal extreme distributions to be GPD, it is possible to create a bivariate equivalent that will give an approximation to the joint distribution. Coles and Tawn (1991, 1994) describe techniques in which the bivariate extremes of the distribution of wave heights and water levels are constructed by fitting a GPD to the marginal distributions. The data are then transformed to normal scales before a bivariate normal model is used to match the dependence in the observations. Hames (2006), Hawkes

(2008), Hames and Reeve (2007), and Galiatsatou and Prinos (2007) provide extensive bibliographies of work on joint probability analysis for coastal engineering applications, together with example applications.

While it is certainly possible to use multivariate extremes in reliability analysis, many of the analytical techniques rely on assumptions of normality and independence between variables. Where numerical methods are required, for example, to solve equations of fluid flow or transmission of forces in structures, it is often more expedient to use either Monte Carlo simulation or Monte Carlo integration techniques, which are discussed further in Chapter 5.

For now, note that to solve a reliability problem it may not actually be necessary to know the joint distribution of the basic variables. The reliability of a structure is often assessed by using the formulae used in design to determine whether extreme conditions will cause a load in excess of a threshold level that is deemed safe. In this case the load may be a pressure, a water level, a flow rate or similar. It will often be a function of a combination of the basic variables, and is sometimes termed the 'structure function'. As an example, consider wave overtopping. Knowing the joint distribution of the wave variables and water levels (four variates) would be helpful but is not necessary. Given a time series of wave conditions and water levels near the toe of a structure, information about the structure geometry and material, and beach levels, it is a simple, but somewhat tedious, process to convert this into a time series of overtopping volumes using a standard design formula for wave overtopping. This is a time series of a single variable, which can be analysed using univariate techniques. At a stroke, all the issues about dependence between variables have been removed. However, the price to pay is twofold. First, the time series of overtopping volumes refers specifically to a particular structure, so that the analysis will have to be run again for each new structure. This is not a major hurdle for coastal structures, because wave conditions can vary quite rapidly along the shoreline and computing design criteria for specific schemes is a commonplace practice. Secondly, the function defining the overtopping volume is highly nonlinear and can have the effect of increasing the scatter in the data, thereby increasing the uncertainty in the estimation of distribution parameters. Similar comments can apply to hydrological, fluvial and estuarine structure functions.

This chapter concludes with a final comment regarding directionality, which applies to winds, currents and waves. Direction as a variate is usually dealt with using one of two methods. Either observations are binned according to direction sectors, with each sector being modelled independently, or the direction is included as a covariate, as in Ewans and Jonathan (2006). Thompson et al. (2008) proposed a method based on Bayesian quantile regression that models the dependence of the extremes of one variable on the value of another using a spline. Inference is performed using a Markov Chain Monte Carlo algorithm. The details of these ideas are beyond the scope of this book, but can be found in Yu and Moyeed (2001) and Yu et al.

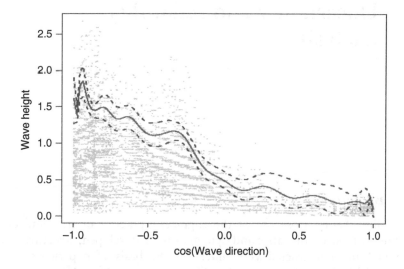

Figure 4.18 A scatter plot of significant wave heights against corresponding transformed wave directions. The waves are hindcast for a point near Selsey Bill from wind records and cover the period between 1971 and 1998 at 3-hourly intervals. The plot includes a 90% Bayesian quantile smoothing spline with 95% credibility envelope. The spline provides a better model, as it is fitted locally, than a single polynomial fitted over the all-direction values (from Thompson 2009).

(2003). A typical output from this type of analysis is shown in Figure 4.18. This shows the best fit to the 90% wave height quantile as a function of wave direction.

Further reading

Beirlant, J., Goegebeur, Y., Segers, J., Teugels, J., de Waal, D. and Ferro, C., 2004. *Statistics of Extremes: Theory and Applications*, John Wiley & Sons Ltd, Chichester, p. 522.

Hawkes, P. J., 2008. Joint probability analysis for estimation of extremes. *Journal of Hydraulic Research*, 46, Extra Issue 2: 246–256.

Kotz, S. and Nadarajah, S., 2000. *Extreme Value Distributions*, Imperial College Press, London, p. 185.

Smith, R. L., 1990. Extreme value theory, in *Handbook of Applicable Mathematics* (Ed. W. Lederman), Vol. 7, John Wiley & Sons, Chichester.

Walshaw, D. and Coles, S. G., 1994. Directional modelling of extreme wind speeds. *Journal of the Royal Statistical Society: Applied Statistics*, part 1, 43: 139–157.

5 Uncertainty and reliability analysis

5.1 Background

Reliability theory arose out of the engineer's practice of trial and error. Before the theories of statics and dynamics were developed, structures were designed and constructed very much on the basis of experience, being informed by the performance of earlier structures. After Newton's work on forces and dynamics, a consistent framework for designing structures evolved that was based on the principles of mechanics. Structures were still not perfect because of several factors. Firstly, the behaviour and performance of materials such as timber, stone and brick depend on their chemical and physical microstructure, which can vary between individual beams or blocks. No two timber beams are identical because the growth rates and internal structure of the fibres differ. Similarly, no two stone blocks are identical at the microscale because the arrangement of the crystalline structure is not repeated exactly from one block to the next. Construction materials have associated with them some uncertainty in their performance due to unavoidable small-scale differences in their composition. Secondly, the diligence with which the construction of a structure follows the design is important. Errors and carelessness in the process of construction can lead to underperformance of the structure. Thirdly, the loads that the structure is intended to withstand may be underestimated. Finally, the prevailing environmental conditions may alter to an extent that diminishes the integrity of the structure. For example, tunnelling beneath buildings can weaken the load-bearing capacity of the earth, leading to subsidence or collapse. Similarly, changes in river flow due to the construction of weirs and dams may lead to higher water levels than previously experienced *and designed for* along some reaches, thereby leading to localised flooding.

The growing reliance on mechanics led towards a somewhat deterministic approach to design. That is to say, it was considered that one could determine exactly the environmental loads that a particular structure was required to resist and then, through knowledge of the behaviour of the construction materials and mechanics, calculate an appropriate sizing of beams, blocks and so on. However, because of the factors mentioned above, this approach did not always meet with success. There was, in fact, if nothing else, an inherent uncertainty in the performance of materials and in the estimation of loads. In due course this led to the realisation that robust design procedures needed to encapsulate these various sources of uncertainty in order to achieve reliable, yet economical, designs.

Reliability theory developed originally as part of structural engineering, and hence much of it is couched in these terms. Thus, when considering the reliability of an object is determined by considering the balance between the 'strength' of the object and the 'load' imposed upon it, the strength and load are denoted by R and S, respectively, by international agreement. The choice of symbols comes from the French words 'résistance' and 'sollicitation' ('asking'). How strength and load are defined will depend on the object or structure in question, and also what mechanisms of failure are being considered.

The theory of reliability is concerned with the definition of the loads and strengths. Their ratio and difference provide means of quantifying how well an object can resist a load. In the following sections these issues are discussed in more detail.

5.2 Definition of failure

Defining what is meant by 'failure' is, it turns out, very important and not always straightforward. We could say a structure fails when it does not perform its intended function. Without defining what the intended function is, this definition is vague at best. As a specific example, consider a flood embankment on a river. We could say that the embankment fails when water passes over the top of the embankment. However, the defence may also 'fail' if there is excessive piping of water through the embankment, excessive scour at the toe, and consequent slumping and thinning of the embankment, loss of cohesion and stability; or from more direct damage, such as by ship collision. Before attempting a reliability analysis, what is meant by failure must be clearly defined.

Types of failure are often split into three categories:

1 Ultimate limit states (ULS) – essentially the point at which the strength equals the load, beyond which there will be damage to the structure and/or reduction in load resistance. Examples of this mode of failure include storm wave conditions in excess of those for which the structure was designed, extreme water levels exceeding embankment levels, and surcharging of drains.
2 Serviceability limit states (SLS) – essentially limits imposed by constraints other than structural integrity. Examples include limitations imposed by gradual deterioration, climate change, environmental and health legislation, maintenance costs or human comfort.

3   Fatigue limit states (FLS) – limits related to loss of strength under repeated loading. This type of failure is not often considered explicitly in the reliability of hydraulic structures due to the nature of the greatest loads (which tend to be intermittent rather than repeated) and the materials used (earth, clay, rock and concrete, rather than steel or plastic). However, fatigue failures have been reported in the prestressing strands of post-tensioned concrete elements. Structures in an intertidal zone will also experience repeated loading due to the regular changes in the ocean tides, and coastal and marine structures will be exposed to repeated loading by waves.

In what follows we will be primarily concerned with ULS, and, to a lesser extent, SLS. Accurate assessment of FLS requires a good understanding of the deterioration of the materials and links between elements of the structure over time. A detailed discussion of this is beyond the scope of this book, although time-dependent risk assessment is touched upon in Section 5.8.

5.3   Uncertainty

In any assessment of strength and loads there will be uncertainties. If the structure under consideration is one that has been made many times before, the historical record of the performance of these structures can be used to build a picture of the reliability. In the case of hydraulic and coastal engineering, many structures are unique in terms of the characteristics of the loading, the materials used and the method of construction.

As already mentioned, due to the variability in timber and rock arising from their natural characteristics, there will be some uncertainty in their response to a specific load. Although similar considerations apply to concrete units, the variability in performance between individual units is generally less in a well-managed operation. This is one reason why concrete units are sometimes favoured over rock, despite the associated visual appearance and energy cost.

Uncertainties also arise in defining load variables. For example, in designing an embankment to resist the 1 in 20 year flood it is useful to have recordings of water levels over at least 20 years, and preferably much, much longer. There are likely to be errors in recording such data, instrument inaccuracies, and even changes in recording behaviour if the method of recording changes (e.g., changing the method of recording from reading levels off a tide board to an automated pressure gauge).

The quality and duration of the observational record also has an influence on any statistical analysis employed to estimate extreme values corresponding to particular return periods. This can affect the quality of both the goodness-of-fit of a particular statistical distribution and any extrapolations made from this.

Failure in the ULS sense is often defined through use of design equations. These are equations based on combinations of experimental observation and mathematical modelling that define a load in terms of basic hydraulic variables and variables describing the strength of the structure. Many modes of failure are described in terms of such design equations. There is, of course, uncertainty in the design equations arising from their empirical nature. Thus, they can only describe behaviour within the range of variables tested in the laboratory. If these experiments were performed using scaled models, there will be uncertainty in the verisimilitude of the scaled observations. Furthermore, there will inevitably be some scatter in the observations, and the design equation will embody a 'best-fit' curve through these points. Figure 5.1 illustrates these different types of uncertainty.

A source of uncertainty that is difficult to quantify is that arising from human errors. Tables 5.1 and 5.2 show the results of studies into the type of human error that can occur and also the relative importance of human error in structural failures.

The clear conclusion from these data is that human error is a significant cause in the majority of recorded failures. Human error must be considered in a reliability assessment. Unfortunately, the means of quantifying human error are limited by our lack of understanding (e.g., Blockley 1992). Another significant underlying cause of failure that is not explicitly identified in the tables is the use of new materials, novel construction methods or construction under circumstances in which the loads are unfamiliar or unknown.

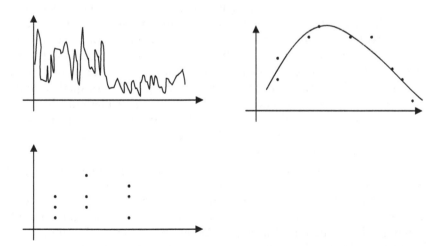

Figure 5.1 Some different forms of uncertainty. Top left – a change in recording device during the record can lead to a change in sensitivity of the measurements and also the mean level; top right – empirical formulae are an idealised fit to discrete measurements; bottom left – repeating the same experiment will often lead to a scatter in the measured values for what should be the same conditions.

Table 5.1 Types of error in observed failures (adapted from Matousek and Schneider 1976)

Factor	Percent
Ignorance, negligence, carelessness	35
Errors, forgetfulness	9
Insufficient control	6
Insufficient knowledge/training	25
Underestimation of conditions	13
Other	12

Table 5.2 Main causes of failure (adapted from Walker 1981)

Cause	Percent
Inadequate appreciation of loads or response	43
Errors in calculations or drawings	7
Insufficient information or contravention of requirements in contract documents	13
Deficiency in construction procedures	13
Random variations in loading, materials and workmanship	10
Other	14

Despite the range of advice and guidance, failures of structures still occur. Such failures can be the result of the structure being exposed to extreme events, greater than the structure was designed to endure, or due to deficiencies in the design or construction. When failures do occur, it is not always clear that there is a single mode of failure. Rather, it may be a combination of factors, such as inappropriate aspects of design, lack of quality control in construction, use of novel construction materials or methods. Even considering a specific example of, say, breaching of an embankment, a single mode of failure may not be easy to identify *a posteriori*. Figure 5.2 shows some failure modes of a shingle bank.

The breaching process may have been initiated by seepage through an internal layer, leading to washout of fines from the core, then to piping, which in turn led to localised slumping, and hence lowering of the crest, resulting in overflow and thence erosion of the crest and backface, and ultimately to the formation of a breach. When investigating the site after the failure it might well be impossible to determine the initiating mechanism. There can, therefore, be significant uncertainty in identifying the mechanism(s) responsible for failure when conducting a forensic assessment, or indeed when trying to understand the risks in the process of design.

A good example of this is shown in Figure 5.3, of flooding that occurred in 2007 near the Blyth Estuary in Suffolk, UK. The footpath lay in a straight

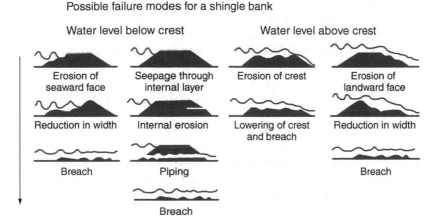

Possible failure modes for a shingle bank

Figure 5.2 Possible failure modes for a shingle bank (after MAFF 2000).

Figure 5.3 Breached embankment and footpath, Bailey Bridge Footpath, Walberswick, Suffolk, UK, arising from a combination of high tides and torrential rainfall in November 2007 (courtesy of Halcrow Group Ltd).

line, in the direction of view, towards the houses in the distance. As well as overflow, evidence of scour on what was the backface of the embankment is evident from the wake pattern in the turbulent flow. Overflow was almost certainly one contributing factor, but seepage and piping may also have combined to lower the crest level at a particular section of the defence. It is difficult to tell.

A question that is almost unanswerable is 'Have all possible modes of failure been considered?'. Sometimes an 'unimaginable' phenomenon occurs that causes structural failure. The Tacoma Narrows Bridge is perhaps one widely-known example of such an instance, where a violent oscillation of the bridge deck was set up by the cross-wind; but the bridge was also of a design that departed from earlier suspension bridge designs. Such 'unimaginable' cases often arise when design techniques or the form of construction generates uncertainty about the performance of the structure under severe conditions. It is of particular importance for innovative projects that attempt to extend the 'state of the art', but which are virtually impossible to quantify in an objective manner. One example of such a case was the \$178 million Sines breakwater in Portugal, which failed catastrophically on 26 February 1978.

The breakwater was constructed in deep water with high wave exposure (the design wave height was $H_s = 11\,\text{m}$), used an armour almost twice as heavy as any used previously, and the exposure and operational techniques made construction difficult. The armour consisted of some 21,000 42-t Dolosse units. Figure 5.4 shows examples of some concrete units used

Figure 5.4 Concrete breakwater armour units (a) Dolosse, (b) Tetrapod, (c) Core-loc and (d) Accropode.

in coastal defence, including Dolosse units. The failure occurred towards the completion of the breakwater. By the time the storm of 26 February 1978 had calmed, almost two-thirds of the Dolosse units in the armour layer were apparently lost (Herzog 1982).

Analysis of the damage of the breakwater (ASCE 1982) showed that units above water level were, by and large, less damaged than those below it. This led to the conclusion that failure occurred due to slumping of the armour units down the seaward slope. What it was not possible to determine was whether the slumping was due to the fracturing of the units, to inaccuracies in placing the units during construction, or to scouring of the seabed at the toe of the structure. The destruction of Sines breakwater also led to numerous studies, amongst which was an investigation of the strength of Dolosse units and the effects of reinforcement (Burcharth et al. 2000). Perhaps the most important conclusion from this failure is that design principles established for smaller structures in shallow water cannot necessarily be transferred directly to larger structures in deep water.

5.4 Risk and reliability

5.4.1 Introduction

The words 'risk' and 'reliability' are well known in everyday parlance. Risk conveys the concept that there is an element of uncertainty, perhaps with the possibility of being reduced if further information is obtained. If something is 'reliable' or has a high degree of reliability then we understand that, unless we are unlucky, the 'thing' will perform its expected purpose. In order to make progress, these ideas have to be defined in a more rigorous manner, using concepts from probability and statistics.

Methods to deal with uncertainty in engineering calculations have been developed. Scientists had already used statistics to describe many natural phenomena, such as water levels, rainfall, wave heights and so on. However, Benjamin and Cornell (1970) demonstrated how statistical techniques could be introduced to engineering calculations so that uncertainties were represented by probability density functions. CIRIA (1977) presented the methods in a widely accessible form. Subsequent reports have specialised the approach for coastal structures, beaches and tidal defences (CIRIA & CUR 1991, CIRIA 1996, EA 2000, Oumeraci et al. 2001).

The application of probabilistic calculations to engineering design, or *probabilistic design* has long been an integral part of design guidance for structures in many countries (e.g., US ACE 1989). The requirement for probabilistic assessments in the design of sea defences was formalised in the UK by MAFF (2000). In The Netherlands, where much of the inhabited land is below mean sea level, probabilistic assessment of coastal defences has been in practice for longer (see e.g., Vrijling 1982).

5.4.2   *Defining risk*

In some texts on structural reliability theory, 'risk' is defined as the probability of failure; that is, the probability of the ultimate limit state being exceeded with a load greater than the strength occurring in any of the possible 'failure modes'. In the context of hydraulic and coastal engineering, it is more conventional to define risk as the combination of the hazard and the consequences of the hazard occurring. That is:

$$Risk = hazard \times consequence \tag{5.1}$$

where the *hazard* is defined as the probability of the load exceeding the strength, and the *consequence* is defined in terms of money or fatalities. The *hazard* is sometimes termed the *probability of failure*. Consequences may be measured in many forms but are often converted to a monetary value, so risk has units of rate of expenditure (e.g., $/quarter, £/year, €/month). In the case of flooding, the *consequences of failure* occurring will depend on the geographical extent of the inundation, its duration and the nature of whom and what are adversely affected. The *risk* of inundation is then the combination of the probability of failure and an evaluation of the consequences. With this definition, it is possible to significantly reduce the *risk* associated with failure simply by reducing the consequences of failure through, for example, setting up a flood-warning system or excluding development in areas prone to flooding, such as river and coastal flood plains.

5.4.3   *Defining reliability*

In everyday parlance something is considered reliable if it works, as noted in Chapter 1. A more formal definition of a reliability function in terms of the cumulative distribution function of a variable was given in Chapter 2. Here, this definition is expanded to the case where there is a function of at least two variables: strength and loading. Strength (R) and loading (S) were introduced in Section 5.1 as a means of assessing the performance of a structure under design conditions. Strength and loading are usually both functions of many variables. The load variables normally include: flow rates (for pipes); water levels (for rivers); wave height, period and direction, and water level (for coasts and estuaries).

The geometry and material of the structure and the characteristics of the riverbed/beach are typical strength variables. When no damage or excess is allowed, the condition $R = S$ is applied. This is known as the 'ultimate limit state condition'. The probability of failure is the probability that the loading exceeds the strength, that is, that $S > R$, and the reliability is then defined as the probability that $S \leq R$.

In the traditional design approach, a limit state condition is set in accordance with the accepted loading of the structure. Exceedance of the limit state condition (i.e., 'failure') is accepted with a small probability P_f. P_f is often expressed as the reciprocal of the return period of exceedance ($P_f = 1/T_R$), where T_R is the return period of the loading in the limit state condition and has the units of a rate. Typically, a probability of failure will be quoted as a percentage, that is, $x\%$, with the implicit time unit understood. For the case where there is a single known load s, the probability of failure is simply $P(R - s \leq 0)$, or $F_R(s)$. Where both the strength and the load are considered as random variables then, if the probability distributions for the strength and loading are $F_R(R)$ and $F_s(S)$, respectively, the probability of failure is given by:

$$P_f = \int_{-\infty}^{\infty} F_R(x) f_S(x) dx \qquad (5.2)$$

under the condition that R and S are independent. This equation is best understood by plotting the density functions of R and S, as shown in Figure 5.5. Equation (5.2) gives the probability of failure as the product of the probabilities of two independent events summed over all possible occurrences. Suppose the loading has a value x. The probability of this occurring is given by $f_S(x)dx$ (i.e., the probability that S lies close to x, within an

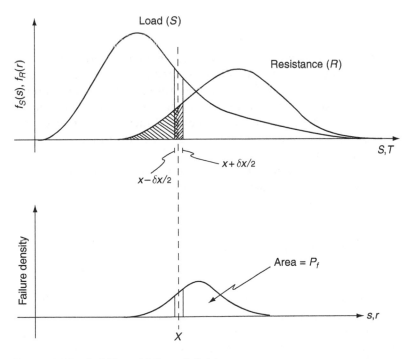

Figure 5.5 Probability of failure definitions.

interval of length dx). Failure will occur if the strength R is less than x. Now, $F_R(x)$ is the probability that the strength R is less than x, so the integrand is the probability that for a given load $S = x$, failure will occur. The total probability of failure is obtained by summing (or, in this case, integrating, as the variables are considered continuous) over all possible values that x can take. As a specific example, consider R to be the level of an embankment that has some uncertainty associated with it and S to be the variable water level. Failure occurs when the water level exceeds the embankment level. To calculate the probability of failure it is necessary to consider each possible water level and the probability that the embankment crest is below this water level.

In general, a reliability function G is defined as

$$G = R - S$$
$$= R(x_1, x_2, \ldots, x_m) - S(x_{m+1}, \ldots, x_n) \tag{5.3}$$

where $x_1, x_2, \ldots, x_m$ are strength variables, $x_{m+1}, \ldots, x_n$ are load variables and the probability of failure (i.e., the probability that $G < 0$) is evaluated from

$$\int_{G<0} f_G(\underline{g}) \, d\underline{g} \tag{5.4}$$

where $\underline{g} = (x_1, x_2, \ldots, x_m, x_{m+1}, \ldots, x_n)$, $f_G(\underline{g})$ is the joint probability function of $\underline{g}$ and the integral is over a volume defined in n dimensions. In practice, a design problem will involve many variables, and the evaluation of the probability of failure will involve integrating over a volume in many dimensions. An additional complication can arise if there is dependence between strength and load variables, such as through the effect that beach levels can have on wave conditions at a structure.

A common assumption in much reliability analysis is to take the variables $x_1, x_2, \ldots, x_n$ to be independent so that Equation (5.4) reduces to a multiple integral

$$\int\int\int_{G \, < \, 0}\int f_{x_1}(x_1) \cdot f_{x_2}(x_2) \cdots\cdots\cdots\cdots f_{x_n}(x_n) \, dx_1 dx_2 \ldots\ldots dx_n \tag{5.5}$$

where $f_{x_1}(x_1)$, $f_{x_2}(x_2), \ldots\ldots, f_{x_n}(x_n)$ are the marginal probability density functions of the loading and strength variables. Even with the assumption of independence, the integral can be very difficult to evaluate. This has prompted the development of various approximate methods, which are discussed in the next section.

5.4.4 Reliability theories

Much of probabilistic reliability theory is concerned with the approximate evaluation of Equation (5.4) or Equation (5.5). Simpler approaches, based on trying to describe some of the uncertainties associated with traditional methods, are included in the hierarchy of methods described here. In particular, the development of partial safety factors represents a robust means of quantifying the uncertainty in quite complicated structures, such as rubble mound breakwaters (see e.g., PIANC 1992). Finally, with the advent of significant computer power, and the development of more sophisticated numerical modelling methods, direct estimation of the probabilities via Monte Carlo simulation is a possibility in some cases.

Reliability methods can be divided into four categories:

Level 0: Traditional methods that use characteristic values of strength and loading.

In the traditional design approach, characteristic values of strength, $\bar{R}$, and load, $\bar{S}$, are used to ensure that $\bar{R}$ is sufficiently greater than $\bar{S}$ to meet the design requirements. Level 0 methods are not discussed further here.

Level 1: Quasi-probabilistic methods, which assign safety factors to each of the variables to account for uncertainty in their value.

Level 2: Probabilistic methods, which approximate the distribution functions of the strength and load variables to estimate Equation (5.4). Level 2 methods have been further subcategorised as first-order risk methods (FORMs), and second-order risk methods (SORMs), depending on the order of the approximation to the reliability function.

Level 3: The most complex probabilistic methods, which estimate Equation (5.4) either directly or through numerical simulation techniques.

The array of methods that are available can appear rather overwhelming initially. To provide some assistance, Appendix B contains flow charts to assist in selecting the appropriate method for the problem in hand. These will not make much sense to the reader until Sections 5.5, 5.6 and 5.7 have been read.

5.5 Level 1 methods

Level 1 methods are design methods in which appropriate measures of structural reliability are provided by the use of partial safety factors that are related to predefined characteristic values of the major loading and structural variables.

Probabilistic design techniques are based on the ultimate limit state, in which G is the reliability function, R is the resistance (or strength) of the structure and S is the design load. For a structure to resist a specified load S then $R \geq S$, or $R = \Gamma S$ where Γ is a number greater than or equal to 1. Γ is the factor of safety against failure, and is included to account for uncertainty. More generally, we may write

$$G = \frac{R}{\Gamma_r} - \Gamma_s S = 0 \tag{5.6}$$

where Γ_r is a safety coefficient relating to the resistance (sometimes called the 'performance factor'), and Γ_s is a safety coefficient relating to the load. The product $\Gamma_r \Gamma_s$ is the (global) factor of safety, Γ. The values of Γ are greater than one, thereby having the effect of reducing the strength and increasing the load. In practical problems there are usually many variables to consider, and partial safety factors for each variable will be specified. Thus, for example, $\Gamma_r = \Gamma_{r1}\Gamma_{r2}\Gamma_{r3}\Gamma_{r4}\Gamma_{r5}$, where the Γ_{ri} are the partial safety factors for five strength variables used in the definition of the particular reliability function.

In Level 1 methods, R and S are assigned characteristic or mean values. The safety factors are normally specified for a discrete set of values of R and S, being based on laboratory or prototype tests. For many types of failure function, the resistance and loading will depend on several variables, say N. Typically, partial safety factors will be tabulated for each variable, and so the global safety factor will be the product of N partial safety factors. In standard structural design, partial safety factors are provided in building codes and the like, and are based on a large body of designs and tests. A similar volume of accurate measurements is not normally available for coastal structures, and hence the level of confidence in partial safety factors has not been as great, although safety factors for Level 1 design may be found in PIANC (1992) and Burcharth and Sorensen (1998) for rubble mounds and vertical breakwaters, respectively.

5.6   Level 2 methods

5.6.1   *Historical background*

Level 2 methods introduce the concept of probability distributions to the calculations. At the beginning of the development of these methods, the load and strength variables were considered to be independent, normally distributed variables. Furthermore, the reliability function was required to be a linear function of the variables. In the case of two variables, S and R are taken as independent, normally distributed variables with known means and standard deviations. Thus, from Equation (3.58) and Example 3.21, the linear reliability function $G = R - S$ is also normally distributed, with the mean and standard deviation given by

$$\mu_G = \mu_R - \mu_S \tag{5.7}$$

$$\sigma_G = \sqrt{\sigma_R^2 + \sigma_S^2} \tag{5.8}$$

The quantity $(G - \mu_G)/\sigma_G$ is unit standard normal, and thus

$$P_f = P\{G \le 0\} = \int_{-\infty}^{0} f_G(x)dx = \Phi\left(\frac{0 - \mu_G}{\sigma_G}\right) \equiv \Phi(-\beta) \tag{5.9}$$

where β is defined as the reliability index; it is also the reciprocal of the coefficient of variation and is the distance of the most probable value of G to the failure point $G = 0$, measured in units of standard deviations of G. This situation is shown in pictorial form in Figure 5.6.

At failure, $G = 0$ and the probability of failure, P_F, is equal to the area of the shaded region in Figure 5.6.

In fact, it is not necessary that R and S are independent. If they are correlated, with correlation coefficient ρ, then Equation (5.7) remains true but σ_G is given by

$$\sigma_G = \sqrt{\sigma_R^2 + \sigma_S^2 - 2\rho\sigma_R\sigma_S} \tag{5.10}$$

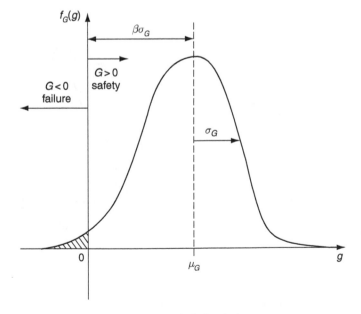

Figure 5.6 Illustration of the reliability index.

For the more general case of a reliability function G that is a *linear* function of n normally distributed variables, Equation (3.62) gives the corresponding expressions for the mean and standard deviation of G. The reliability index β may be given a simple geometric interpretation. Consider the standardised variables

$$R' = \frac{R - \mu_R}{\sigma_R} \text{ and } S' = \frac{S - \mu_S}{\sigma_S}$$

In terms of R' and S' the reliability function becomes

$$G = R - S = \sigma_R R' - \sigma_S S' + (\mu_R - \mu_S) \tag{5.11}$$

For $G = 0$ this equation describes a line in the plane, as shown in Figure 5.7.

The shortest distance from the origin to the failure 'surface' is equal to the reliability index.

Example 5.1. The crest level of an embankment over a reach is described by a normal distribution with mean of 5 and standard deviation of 0.5. This is often written as N(5, 0.5). Monthly maximum water levels along the reach obey N(3, 1). What is the probability of flooding?

Solution. Flooding occurs when water level > crest level, or $WL > CL$. So the reliability function can be written as

$$G = CL - WL$$

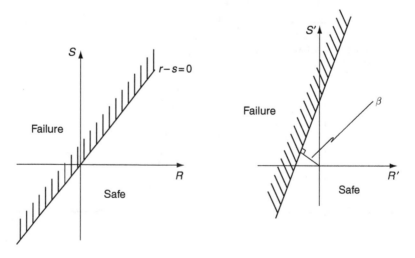

Figure 5.7 Geometric interpretation of the reliability index.

The variables are normal and independent, so (from Equations 5.7 and 5.8):

$$\mu_G = 5 - 3 = 2, \quad \sigma_G^2 = 0.5^2 + 1^2 = 1.25 \Rightarrow \beta = \frac{2}{\sqrt{1.25}}$$

From Equation (5.9),

$$P_F = \Phi \left(\frac{0 - 2}{\sqrt{1.25}} \right) = \Phi(-1.79) = 0.037 \approx 4\%$$

Thus, the probability of failure is approximately 4% per month.

The bulk of Level 2 methods deal with approximations that attempt to reduce the reliability function and the main variables to the simple case of a reliability function that is linear in the variables, and the variables are independent and normally distributed, so that Equations (5.7–5.9) can be applied.

Dealing first with the case where the reliability function is not a linear function of the variables; what can be done? The function can be linearised by expanding it as a Taylor series and truncating quadratic terms and higher. This sounds like a drastic action, but can give useful results under the right conditions. Let

$$\begin{aligned} G &= R(x_1, x_2, x_3, \ldots, x_m) - S(x_{m+1}, x_{m+2}, \ldots, x_n) \\ &= G(x_1, x_2, \ldots, x_n) \end{aligned} \tag{5.12}$$

Expanding this function in a Taylor series about the point

$$(x_1, x_2, \ldots, x_n) = (X_1, X_2, \ldots, X_n) \tag{5.13}$$

and retaining only linear terms gives

$$G(x_1, x_2, \ldots, x_n) \approx G(X_1, X_2, \ldots, X_n) + \sum_{i=1}^{n} \frac{\partial G}{\partial x_i} (x_i - X_i)$$

where $\frac{\partial G}{\partial x_1}$ is evaluated at $(X_1, X_2, \ldots, X_n)$. Approximate values of μ_G and σ_G are obtained from, (see Equation 3.63)

$$\mu_G \approx G(X_1, X_2, \ldots \ldots X_n) \tag{5.14}$$

and

$$\sigma_G^2 \approx \sum_{i=1}^{n} \sum_{j=1}^{n} \frac{\partial G}{\partial x_i} \frac{\partial G}{\partial x_j} \cdot \text{cov}(x_i, x_j) \tag{5.15}$$

If the variables are uncorrelated then the covariances are zero for $i \neq j$ and unity for $i = j$, so that

$$\sigma_G^2 = \sum_{i=1}^{n} \left(\frac{\partial G}{\partial x_i} \right)^2 \sigma_{x_i}^2 \tag{5.16}$$

The quantities $\left(\frac{\partial G}{\partial x} \sigma_{x_i} \right)^2$ are termed the 'influence factors' as they represent the contributions to the variance in G from each individual random variable X_i.

It is worth noting that, in the case where G is a linear function,

$$G(X_1, X_2, \ldots X_N) = a_0 + \sum_{i=1}^{N} a_i X_i$$

We obtain

$$\beta = \frac{a_0 + \sum_{i=1}^{N} a_i \mu_i}{\sqrt{\sum_{i=1}^{N} \sum_{j=1}^{N} a_i a_j \rho_{ij} \sigma_i \sigma_j}} \tag{5.17}$$

Three variants of Level 2 methods that are in current use are now described.

5.6.2   *First-order mean value approach*

This is sometimes denoted by FMA or MVA, for obvious reasons. In this case $(X_1, X_2, \ldots, X_n) = (\mu_x, \mu_{xz}, \ldots . \mu_{xn})$. That is, the failure function is expanded about the mean values of the basic variables. The mean and standard deviation of the reliability function can then be evaluated directly from the equations above.

Example 5.2. A reliability function is given by $G = aX_1^2 - bX_2$, where a and b are positive constants.

a   Use this to calculate the probability of failure, in terms of a and b, with the MVA method, given that $X_1 = N(4, 1)$ and $X_2 = N(1, 3)$.

b   What condition applies to the values of a and b if the probability of failure is to be no greater than 0.023?

c   If now you are told that results from experiments suggest that $b \approx 0.2$, and the manufacturer of the structure asserts that $a = 0.5$, what do you conclude?

Solution.

a   The first step is to calculate the partial derivatives. This gives:

$$\frac{\partial G}{\partial X_1} = 2aX_1$$

$$\frac{\partial G}{\partial X_2} = -b$$

Table 5.3 summarises the results of the MVA calculations. Thus, we have:

$$\mu_G = \mu_{X_1} - \mu_{X_2} = 3$$

and

$$\sigma_G^2 = 64a^2 + 9b^2$$

Therefore the reliability index $\beta = \frac{\mu_G}{\sigma_G} = \frac{3}{\sqrt{64a^2 + 9b^2}}$
and the probability of failure is given by $\Phi(-\beta)$

b   Now, from Table D in Appendix A, $\Phi(-2.0) = 0.023$, so β must be no smaller than 2 in order for the probability to be no greater than 0.023. Thus, the condition on a and b is:

$$\frac{3}{\sqrt{64a^2 + 9b^2}} \geq 2$$
$$\Rightarrow \quad 256a^2 \leq 9 - 36b^2$$

Now, the left-hand side of this inequality is non-negative, so the right-hand side must also be non-negative. This gives the further constraint on b: $b \leq 1/2$.

Table 5.3 Summary of the steps in the MVA

Variable	Mean	Standard deviation	Partial derivative (evaluated at the mean values)	Influence factor
X_1	4	1	$8a$	$64a^2$
X_2	1	3	$-b$	$9b^2$

c   Using the values of a and b provided, we see that

$$256a^2 = 64$$

and

$$9 - 36b^2 = 7.56 < 64$$

The condition on a and b is not fulfilled, so the desired probability of failure cannot be achieved.

The MVA method is relatively easy to use but can be inaccurate if the failure function is strongly nonlinear. The method also relies on accurate estimates of the mean and variance of the key variables. In the example above, these values have been assumed to be known exactly, but in practice there is likely to be considerable uncertainty in estimating both the mean and the variance. Where parameters have been measured in a series of experiments, the sample mean and variance could be used in the absence of other information. Experience shows that this approach should not be used in isolation. If in doubt, results should be checked against other methods.

5.6.3   *First-order design point approach*

A serious objection to the MVA method is that the point about which the failure function is linearised is not necessarily on the failure surface. Furthermore, the value of the reliability index can change when different, but equivalent, non-linear failure functions are used. A means of overcoming these problems was introduced by Hasofer and Lind (1974), who defined a modified form of reliability index based on expanding about a point in the failure surface. In this first-order design point approach (FDA), commences, as before, with a failure function that is a function of normal, independent, random variables, $x_1, x_2, \ldots, x_n$. The first step is to map these into standard form by

$$z_i = \frac{x_i - \mu_{xi}}{\sigma_{x_i}} \qquad\qquad i = 1, 2, \ldots, n \qquad\qquad (5.18)$$

so that $\mu_{z_i} = 0$ and $\sigma_{z_i} = 1$

Note that the reliability function $G(x_1, x_2, \ldots, x_n)$ is also transformed to a (different) function of the standardised variables, which is denoted by $g(z_1, z_2, \ldots, z_n)$. Hasofer and Lind's reliability index is defined as the shortest distance from the origin to the failure surface in the standardised z-coordinate system. This is shown for two dimensions in Figure 5.8.

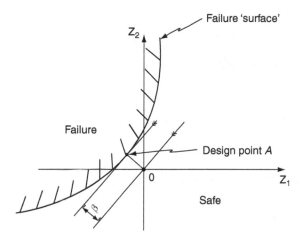

Figure 5.8 Geometric interpretation of Hasofer and Lind's reliability index.

The point A is known as the design point. In general, the reliability index can be found from

$$\beta_{HL} = \min \left(\sum_{i=1}^{n} z_i^2 \right)^{\frac{1}{2}} \tag{5.19}$$

for z_i in the failure surface. The calculation of β_{HL} may be performed in several ways, but generally involves iteration, which is suited to numerical schemes (see e.g., Thoft-Christensen and Baker 1982, Kottegoda and Rosso 1997). The important feature of Hasofer and Lind's reliability index is that it relates to the failure 'surface' $g(z_1, z_2, \ldots, z_n) = 0$, which is invariant with respect to the reliability function because equivalent reliability functions result in the same failure surface. The two reliability indices will match when the failure surfaces are linear. Clearly, this is also the case if nonlinear reliability functions are linearised by Taylor series expansion around the design point. Linearisation about the design point rather than mean values is thus preferable, but has the disadvantage of requiring extra calculations. Linearisation about mean values can give wildly inaccurate results, but because of the simplicity of the evaluation might be used to obtain an initial 'order of magnitude' estimate of the probability of failure.

Numerical solution

In the iterative numerical solution scheme, an initial guess for the design point is chosen, and then a sequence of calculations is performed to yield a refined estimate for the design point. In essence, the procedure 'nudges' our

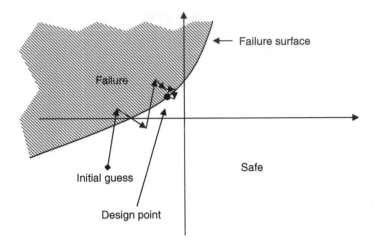

Figure 5.9 Iterative refinement to obtain the design point (in two dimensions).

chosen point, so that not only does it lie on the failure surface but also it is the point on the failure surface that is closest to the origin. The procedure is strictly a minimisation problem. There are several ways in which a solution can be found (see Melchers 1999). Here, we outline a procedure that is based on iterating the initial guess along the failure surface to the point at which the normal to the surface passes through the origin. This is illustrated in Figure 5.9.

Let x denote the basic variables $x_1, x_2, \ldots, x_n$ and z denote the standardised variables $z_1, z_2, \ldots, z_n$. Furthermore, let $z^{(m)}$ denote the mth approximation to the point representing the line from the origin to the failure surface $g(z) = 0$. To create an iterative scheme requires an expression relating successive approximations $z^{(m)}$ and $z^{(m+1)}$. This can be obtained from a first-order Taylor expansion of $g[z^{(m+1)}]$ about $z^{(m)}$:

$$g[z_1^{(m+1)}, z_2^{(m+1)}, \ldots, z_n^{(m+1)}] \approx g[z_1^{(m)}, z_2^{(m)}, \ldots, z_n^{(m)}]$$

$$+ \sum_{i=1}^{n} (z_i^{(m+1)} - z_i^{(m)}) \frac{\partial g(z^{(m)})}{\partial z_i} = 0 \qquad (5.20)$$

In the language of mathematics, this expression represents a hyperplane, $g[z^{(m+1)}] = 0$, approximating the hypersurface $g[z^{(m)}] = 0$ in n-dimensional z-space, for which at point $m + 1$, the linearised reliability function must be satisfied. In two dimensions this is equivalent to defining the tangent to a curve, and in three dimensions to defining the tangent plane at a point on a solid object.

From the geometry of surfaces (e.g., Lang 1974) the outward normal vector to a hyperplane has components given by

$$c_i = \lambda \frac{\partial g}{\partial z_i} \tag{5.21}$$

where λ is an arbitrary constant. The total length of the outward normal, l, is defined as

$$l = \left(\sum_i c_i^2 \right)^{1/2} \tag{5.22}$$

and the direction cosines, α_i, of the unit outward normal are then

$$\alpha_i = \frac{c_i}{l} \tag{5.23}$$

With α_i known, it may be shown that the coordinates of the trial point $z^{(m)}$ are

$$z^{(m)} = -\alpha^{(m)} \beta^{(m)} \tag{5.24}$$

Substituting Equation (5.24) into Equation (5.20) and rearranging yields:

$$z^{(m+1)} = -\alpha^{(m)} \left[\beta^{(m)} + \frac{g(z^{(m)})}{l} \right] \tag{5.25}$$

The 'recipe' for numerical solution can now be stated as:

1   Standardise the basic random variables x to the independent standardised normal variables z.
2   Transform $G(x)$ to $g(z)$.
3   Choose initial guess $z^{(1)}$.
4   Compute $\beta^{(1)} = \left[\sum_{i=1}^{n} \left(z_i^{(1)} \right)^2 \right]^{\frac{1}{2}}$; set $m = 1$.
5   Compute direction cosines $\alpha^{(m)}$ using Equations (5.21–5.23).
6   Compute $g[z^{(m)}]$.
7   Compute $z^{(m+1)}$ using Equation (5.25).
8   Compute $\beta^{(m+1)} = \left[\sum_{i=1}^{n} \left(z_i^{(m+1)} \right)^2 \right]^{\frac{1}{2}}$.
9   Check whether $z^{(m+1)}$ and $\beta^{(m+1)}$ have converged; if not, go to step (5) and increase m by 1.
10 Once $z^{(m+1)}$ and $\beta^{(m+1)}$ have converged, use Equation (5.9) to evaluate the probability of failure.

It should be noted that this procedure is not guaranteed to converge nor, if it does converge, to converge to the global minimum value of β. To alleviate this problem it is advisable to repeat the calculations several times with different initial guesses, to check that the solution converges to the same endpoint in each case.

One (of many) alternative iterative schemes is based on the following. Let z^* be the failure point. By careful examination of Equations (5.21)–(5.25) it can be seen that any component of this, z_i^*, is given by

$$z_i^* = \frac{\left(\dfrac{\partial g}{\partial z_i}\right)\Big|_{z=z^*}}{\sqrt{\displaystyle\sum_{i=1}^{n}\left(\dfrac{\partial g}{\partial z_i}\right)^2\Big|_{z=z^*}}}\beta = -\alpha_i\beta \tag{5.26}$$

where we have defined

$$\alpha_i = \frac{\left(\dfrac{\partial g}{\partial z_i}\right)\Big|_{z=z^*}}{\sqrt{\displaystyle\sum_{i=1}^{n}\left(\dfrac{\partial g}{\partial z_i}\right)\Big|_{z=z^*}}} \tag{5.27}$$

or,

$$\alpha_i = \frac{\left(\dfrac{\partial g}{\partial z_i}\right)\Big|_{z=z^*}}{K} \tag{5.28}$$

where

$$K = \left[\sum_{i=1}^{n}\left(\frac{\partial g}{\partial z_i}\right)^2\right]^{1/2} \tag{5.29}$$

By definition,

$$x_i^* = \mu_i + \sigma_i z_i^* = \mu_i - \alpha_i\sigma_i\beta \tag{5.30}$$

The required value of β can then be found by substituting this expression for the x_i in the reliability function G:

$$G(\mu_1 - \alpha_1\sigma_1\beta, \mu_2 - \alpha_2\sigma_2\beta, \ldots, \mu_n - \alpha_n\sigma_n\beta) = 0 \tag{5.31}$$

An iterative procedure is then launched with an initial guess for z. Then, using estimated or known values of the means and variances, one determines the x_i values, the partial derivatives, and the values of α_i, which are then used in Equation (5.25) to solve for β. This value of β is then used in Equation (5.30) to provide a refined estimate of x_i^*. The process is repeated until convergence.

Correlated normal variables

The case where the basic variables are normal and correlated can be treated with a slight modification of the above method. The correlation between any pair of the random variables **x** is expressed by the covariance matrix

$$\text{Cov}_x = \begin{bmatrix} \text{Var}(x_1) & \text{Cov}(x_1,x_2) & \cdots & \text{Cov}(x_1,x_n) \\ \text{Cov}(x_2,x_1) & \text{Var}(x_2) & & \vdots \\ \vdots & & \ddots & \vdots \\ \text{Cov}(x_n,x_1) & \cdots & \cdots & \text{Var}(x_n) \end{bmatrix}$$

The covariance matrix is symmetric [i.e., $\text{Cov}(x_i,x_j) = \text{Cov}(x_j,x_i)$] and positive definite. It is well known from linear algebra that a real-valued, symmetric matrix may be transformed into a diagonal matrix, that is, a matrix whose elements are all zero except those on the diagonal. This process corresponds to a transformation from $x_1,x_2,\ldots,x_n$ to $y_1,y_2,\ldots,y_n$, where each y_i is a linear combination of the x_i' values, and the y_i values are independent normal variables. The transformation is determined by finding the eigenvectors of the covariance matrix Cov_x. The eigenvalues of Cov_x correspond to the variances of the y_1 values. Thus,

$$\begin{bmatrix} y_1 \\ y_2 \\ \vdots \\ y_n \end{bmatrix} = \begin{bmatrix} a_{11} & a_{21} & \cdots & a_{n1} \\ a_{12} & a_{22} & & \vdots \\ \vdots & & \ddots & \vdots \\ a_{1n} & \cdots & \cdots & a_{nn} \end{bmatrix} \begin{bmatrix} x_1 \\ x_2 \\ \vdots \\ x_n \end{bmatrix}$$

and

$$\text{Cov}_y = \begin{bmatrix} \text{Var}(y_1) & 0 & \cdots & 0 \\ 0 & \text{Var}(y_2) & & \vdots \\ \vdots & & \ddots & \vdots \\ 0 & \cdots & \cdots & \text{Var}(y_n) \end{bmatrix}$$

In general, the eigenvectors and eigenvalues have to be calculated numerically, and techniques for this are well known (e.g., Press et al. 2007). Once the elements a_{ij} and the variances of the y'_i values have been determined in this fashion, it remains to:

1   Write the x_1 values in terms of the y_i values.
2   Write the reliability function $G(x)$ in terms of the y_1 values.
3   Determine the mean value of each y_i from the knowledge of the transformation and the means of the x_1 values.
4   Proceed as in the above 'recipe', using y in place of x.

5.6.4   *Approximate full distribution approach (AFDA)*

In the approximate full distribution approach (AFDA) the FDA is used but the variables are allowed to be non-normal. Equating normal and non-normal distribution functions at the design point allows the requirement for normal variables of the FDA method to be relaxed. This is sometimes referred to as the 'normal tail transform'. Specifically, if the reliability function depends on a non-normal variable Y, this can be rewritten in terms of normal variables through the transformation.

$$Z = \Phi^{-1}[F_Y(y)] \tag{5.32}$$

where $F_Y(y)$ is the distribution function of Y and Φ^{-1} is the inverse normal distribution function. This transformation is shown pictorially in Figure 5.10.

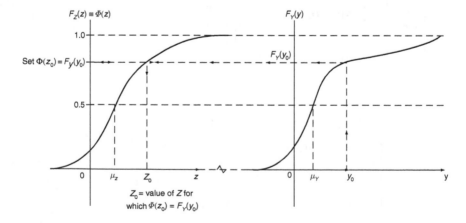

Figure 5.10 The transformation of non-normal variables to equivalent normal variable at the design point.

If the design point is given by $x^* = (x_1{}^*, x_2{}^*, \ldots, x_n{}^*)$, then the transformation reads

$$F_{X_i}(x_i^*) = \Phi\left(\frac{x_i^* - \mu_{X_i}}{\sigma_{X_i}}\right) \tag{5.33}$$

$$f_{X_i}(x_i^*) = \frac{1}{\sigma_{X_i}}\phi\left(\frac{x_i^* - \mu_{X_i}}{\sigma_{X_i}}\right) \tag{5.34}$$

where μ_{X_i} and σ_{X_i} are the mean and standard deviation of the approximating normal distribution. Now,

$$F_{X_i}(x_i^*) = \Phi\left(\frac{x_i^* - \mu_{X_i}}{\sigma_{X_i}}\right) = \Phi(z_i^*) = \Phi(\beta\alpha_i) \tag{5.35}$$

Solving for x_i^* yields

$$x_i^* = F_{X_i}^{-1}\Phi(\beta\alpha_i) \tag{5.36}$$

The iterative method (Equations 5.26–5.31) can be used for this case, as long as the mean and standard deviation obtained by solving Equations (5.33) and (5.34) are calculated for those variables where the normal tail transformation has been applied.

5.6.5 Correlated non-normal variables

Consider a set of n correlated variables $X_1, X_2, \ldots, X_n$. They can be represented only through their joint distribution function, $F_{\underline{x}}(\underline{x})$

$$F_{\underline{x}}(\underline{x}) = P\left[(X_1 \leq x_1) \text{ and } (X_2 \leq x_2) \text{ and } \ldots \text{ and } (X_n \leq x_n)\right] \tag{5.37}$$

However, in many situations, sufficient data and other information are not available to determine the form of the joint distribution function with any certainty. Often, the most that can be expected is that the marginal distributions $F_{X_i}(x_i)$ can be determined, together with their correlation matrix.

For a pair of jointly distributed random variables X_1, X_2, the marginal distribution function is defined as

$$F_{X_1}(x_1) = \int_{-\infty}^{x_1} f_{X_1}(t)dt = \int_{-\infty}^{x_1}\int_{-\infty}^{\infty} f_{X_1, X_2}(t, x_2)dx_2 dt \tag{5.38}$$

and analogously for n jointly distributed variables, where $f_{X_1}(t)$ is the density function of X_1 and so on.

The correlation matrix $\underline{R}$ is given by

$$\underline{R} = \begin{pmatrix} \rho_{11} & \rho_{12} & \cdot & \cdot & \rho_{1n} \\ \rho_{21} & \rho_{22} & \cdot & & \cdot \\ \rho_{31} & \cdot & \rho_{33} & \cdot & \cdot \\ \cdot & \cdot & \cdot & \cdot & \cdot \\ \rho_{n1} & \cdot & \cdot & \cdot & \rho_{nn} \end{pmatrix} \tag{5.39}$$

where ρ_{ij} is the correlation coefficient between variables X_i and X_j with $i,j = 1,2,\ldots,n$. However, if $F_{\underline{x}}(\cdot)$ or $f_{\underline{x}}(\cdot)$ are not known, the marginal distributions can not be obtained from Equation (5.38), and these plus the correlation matrix must be obtained directly by fitting from data.

In the event that the individual marginal distributions of the basic variables $\underline{X}$ and their correlation matrix $\underline{R}$ can be estimated, the method proposed by Der Kiureghian and Liu (1986) may be used. An outline of the method is given here.

A joint density function is assumed that is consistent with the known marginal distributions and correlation matrix:

$$f_{\underline{x}}(\underline{x}) = \phi_n(y, \underline{R}') \; \frac{f_{x_1}(x_1) f_{x_2}(x_2) \ldots f_{x_n}(x_n)}{\phi(y_1) \phi(y_2) \ldots \phi(y_n)} \tag{5.40}$$

where $\phi_n(y, \underline{R}')$ is the n-dimensional normal probability density function with zero means, unit standard deviations and with correlation matrix $\underline{R}'$.

$\phi(y_i)$ is the standard univariate normal density function
$f_{X_i}(x_i)$ is the marginal density function for basic variable X_i.

The elements ρ_{ij} of the correlation matrix R' are related to the known marginal densities $f_{X_i}(x_i)$ and $f_{X_j}(x_j)$ and the correlation coefficient ρ_{ij} between the basic variables X_i and X_j through the relationship

$$\rho_{ij} = \int_{-\infty}^{\infty} \int_{-\infty}^{\infty} \left(\frac{x_i - \mu_i}{\sigma_i} \right) \left(\frac{x_y - \mu_j}{\sigma_j} \right) \phi_2(y_i, y_j.\rho_{ij}') \, dy_i \, dy_j \tag{5.41}$$

where $\phi_2(\cdot, \cdot,)$ is the bivariate normal density with correlation coefficient ρ_{ij},, and x_i is given by

$$x_i = F_{x_i}^{-1}[\Phi(y_i)] \tag{5.42}$$

Solution of Equation (5.41) usually has to be performed iteratively to determine ρ_{ij}', and needs to be solved for each pair of values of i and j.

Once the new correlation matrix $\underline{R}'$ has been obtained, the steps are as follows:

1   Obtain a set of correlated normal variables with zero mean and unit standard deviation y_i from

$$\underline{Y} = \Phi^{-1}\left[F_{\underline{x}}(\underline{x})\right] \qquad \underline{Y} = (Y_1, Y_2, \ldots, Y_n) \qquad (5.43)$$

2   Compute $\underline{L}$ from

$$\underline{R}' = \underline{L}\,\underline{L}^T \qquad (5.44)$$

3   Obtain a set of uncorrelated unit standard normal variables by

$$\underline{Z} = \underline{L}^{-1}\underline{Y} \qquad (5.45)$$

4   The set of correlated, non-normal basic variables $\underline{X}$ occurring in a reliability function may be replaced by

$$X_i = F_{Y_i}^{-1}\left[\Phi(Y_i)\right] \quad i = 1, 2, \ldots, n \qquad (5.46)$$

with Y_i being obtained from Equation (5.43).

Example 5.3. Given that: X_1 is normal with mean $\mu_{x_1} = 100$ and standard deviation $\sigma_{x_1} = 15$, X_2 is Gumbel with mean $\mu_{x_2} = 5$ and standard deviation $\sigma_{x_2} = 1$, their correlation $\rho_{12} = 0.6$, and the reliability function is of the form $G = X_1 - 1.5X_2$, determine the reliability function in terms of uncorrelated normal variables.

Solution.

We have

$$\underline{R} = \begin{pmatrix} 1 & 0.6 \\ 0.6 & 1 \end{pmatrix}$$

and

$$F_{X_2}(x_2) = \exp\left[-\exp\left(-\alpha(x_2 - u)\right)\right]$$

where

$$\alpha = \pi/(\sigma_{x_2}\sqrt{6}) = 1.283$$

$$u = \mu_{x_2} - \frac{\gamma}{\alpha} = 4.55$$

from the properties of the Gumbel distribution

This gives

$$F_{X_2}(x_2) = \exp\left[-\exp\left(-1.283(x_2 - 4.55)\right)\right]$$

Or

$$x_2 = 4.55 - \ln x_2 = 4.55 - \ln[-\ln(F_{X_2}(x_2))]/1.283$$

Therefore

$$\begin{aligned}
\rho_{12} = 0.6 &= \int_{-\infty}^{\infty} \int_{-\infty}^{\infty} \left(\frac{x_1 - \mu_{x_1}}{\sigma_{x_1}}\right) \left(\frac{x_2 - \mu_{x_2}}{\sigma_{x_2}}\right) \Phi_2(y_1, y_2; \rho'_{ij}) \, dy_1 \, dy_2 \\
&= \int_{-\infty}^{\infty} \int_{-\infty}^{\infty} y_1 \cdot \left\{\frac{\{4.55 - [\ln(-\ln(\Phi(y_2)))/1.283]\} - 5.0}{1}\right\} \Phi_2(y_1, y_2; \rho'_{ij}) \, dy_1 \, dy_2 \\
&= \int_{-\infty}^{\infty} \int_{-\infty}^{\infty} y_1 \cdot \{-0.45 - \ln[-\ln(\Phi(y_2))]/1.283\} \Phi_2(y_1, y_2; \rho'_{ij}) \, dy_1 \, dy_2
\end{aligned}$$

Iterative solution of the above integral gives $\rho'_{12} \approx 0.62$ by numerical integration.

Hence, $\underline{R}' = \begin{bmatrix} 1 & 0.62 \\ 0.62 & 1 \end{bmatrix}$

Cholesky decomposition (determined numerically) gives

$$\underline{L} = \begin{bmatrix} 1 & 0 \\ 0.62 & 0.79 \end{bmatrix} \text{ and } \underline{Y} = \underline{L}\,\underline{Z} = \begin{bmatrix} 1 & 0 \\ 0.62 & 0.79 \end{bmatrix} \begin{bmatrix} Z_1 \\ Z_2 \end{bmatrix}$$

Therefore $X_1 = F_{x_1}^{-1}[\Phi(Z_1)] = \mu_{x_1} + Z_1 \sigma_1 = 100 + 15Z_1$ (5.47)

and $X_2 = F_{x_2}^{-1}[\Phi(0.62Z_1 + 0.79Z_2)]$

$$= u - \ln[-\ln(\Phi(0.62Z_1 + 0.79Z_2))]/\alpha$$

$$= 4.55 - [\ln(-\ln(\Phi(0.62Z_1 + 0.79Z_2)))/1.283] \tag{5.48}$$

Finally, $G(X_1, X_2) = X_1 - 1.5X_2$ becomes

$$\begin{aligned}
g(Z_1, Z_2) = {}&100 + 15Z_1 - (1.5)4.55 + (1.5) \\
&\ln[-\ln(\Phi(0.62Z_1 + 0.79Z_2))]/1.283
\end{aligned}$$

i.e.,

$$g = 93.18 + 15Z_1 + [1.5\ln(-\ln(\Phi(0.62Z_1 + 0.79Z_2)))/1.283] \tag{5.49}$$

This reliability function is a (nonlinear) function of independent normal variables Z_1 and Z_2.

5.6.6 Correlated extreme variables

It may have struck the reader that much of the effort in earlier chapters was concerned with determining the distribution function of extreme values. This is because most engineering structures are designed to withstand 'everyday' loads with ease and also to resist unusual loading up to a given limit. We saw in Chapter 4 that the distribution of the extremes of a normal random variable do *not* follow a normal distribution. Furthermore, there may be a complicated dependence structure between joint extremes that is different from that for more 'typical' values of the variates. The methods given in the previous sections can be applied to such cases, but it is as well to remember that extreme distributions are skewed, and that the observational basis for univariate extreme values in hydraulic and coastal engineering is scarce, and for multivariate extremes even more so.

The uncertainties involved in determining a probability of failure of a structure may seem overwhelming, but a pragmatic approach coupled with the foregoing methods can provide useful results. One source of assistance comes from the way in which the strength of hydraulic structures is defined. Very often, this will not be a complicated function of many variables, but simply a number. For example, if we consider overflow of an embankment, then the 'strength' of the structure is just the crest level. To make it slightly more realistic, we might make allowances for local variability in the crest level by assigning a narrow probability distribution function to it, say a N(design crest level, 0.1 m), based on the contractor's estimate of working accuracy or a post-construction survey.

In contrast, much of the complication arises from specifying the loads, which can involve functions of many variables such as head, flow rate, flow velocity, wave height, wave period, rainfall intensity, storm duration and so on. Sometimes, as alluded to in Chapter 4, it may be possible to cast the reliability problem as a function of one or two variables only. As an example, consider wave overtopping of a sea wall. The 'strength' is usually defined in terms of a maximum permissible overtopping rate. The load is defined by an empirical formula involving at least five variables describing the environmental conditions and the sea-wall geometry and materials. Simply determining the joint probability function of these variables would be challenging enough in a standard case. However, if long time series of waves, water levels, beach conditions and so on are available, one alternative is to use these to generate a time series of overtopping rates. These overtopping rates can then be analysed in the standard univariate manner (i.e., using annual maxima or peaks over thresholds) to yield a 'best-fit' probability distribution function for the extreme events, and then use this to calculate the probability of failure – that is, the annual probability that the overtopping at the sea wall will exceed the permissible rate.

Such an approach requires a significant amount of data that will not necessarily be available in all cases, but it allows the complications associated

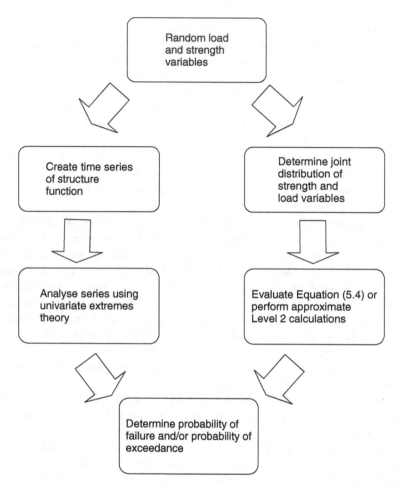

Figure 5.11 Alternative routes to evaluating the probability of failure in the multivariate case.

with potentially correlated, non-normal variables to be circumvented. It is sometimes known as the 'structure function' approach, for obvious reasons. Figure 5.11 illustrates the two alternative routes for calculating the probability of failure in this case.

5.6.7 Point estimation methods

Point estimation methods (PEMs) offer one alternative to the first-order reliability methods described in the previous sections. PEMs, as developed by Rosenblueth (1975) and Li (1992), can be applied to the failure function G to obtain estimates of the moments of the distribution function of G.

The two of interest, the mean and variance, can then be used to evaluate the reliability index, and thus the probability of failure. Alternatively, the estimated moments can be used to define a (chosen) statistical distribution function to G, from which the probability of failure can be determined directly. This method does not require the evaluation of derivatives of the reliability function, which can be advantageous if the reliability function is not given explicitly in functional form, but perhaps instead as implicit functions of the variables, in the form of tables or graphs, or even as the output of a detailed numerical model.

The method requires the statistical moments of the basic variables and their correlations to be known and also for the function G to be evaluated at a specific set of values of the basic variables. Explicit formulae for functions of up to three correlated variables were derived by Rosenblueth (1975). The method assumes, in effect, that the probability mass can be concentrated at a finite number of points. The general method is outlined below.

Let $G = G(X_1, X_2, \ldots, X_n)$ be a reliability function that is a function of n basic variables. Point estimates at 2^n points g_k (with $k = 1, 2, \ldots, 2^n$) are computed as

$$g_k = G(\mu_1 + \varepsilon_{1k}\sigma_1, \mu_2 + \varepsilon_{2k}\sigma_2, \ldots, \mu_n + \varepsilon_{1n}\sigma_n) \tag{5.50}$$

where μ_i and σ_i are the mean and standard deviation of X_i, respectively. The ε_{ik} are coefficients that take the values ± 1, satisfying

$$k = 1 + \sum_{i=1}^{n} 2^{i-2}(1 + \varepsilon_{ik}) \tag{5.51}$$

Thus, if G is a function of a single variable there are two point estimates; if G is a function of two variables there are four point estimates; and so on. The moments of G are estimated from

$$E(G^r) = \sum_{k=1}^{2^n} \theta_k g_k^r \tag{5.52}$$

where

$$\theta_k = \begin{cases} 2^{-n} & \forall k \text{ if } X_i \text{ are mutually independent} \\ 2^{-n}\left(\sum_{i=1}^{n-1}\sum_{j=i+1}^{n} \varepsilon_{ij}\varepsilon_{jk}\rho_{ij}\right) & \text{if } X_i \text{ are correlated} \end{cases} \tag{5.53}$$

and ρ_{ij} is the correlation coefficient between X_i and X_j.

The method is illustrated in Figure 5.12 for the case of two variables.

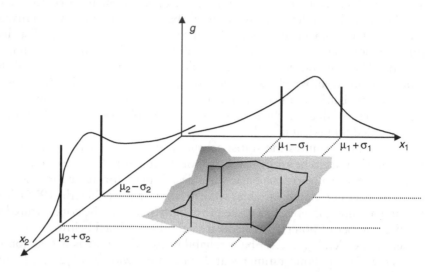

Figure 5.12 Illustration of the point estimation method for a function of two variables, $g(x_1, x_2)$. The moments of g are estimated from the values of g at four points.

The reliability of this method can be impaired if the distribution of G is highly skewed. Modifications of the method have been proposed by Harr (1987) and Chang et al. (1993), amongst others.

As noted above, if there is reason to suspect that the reliability function obeys a particular form of distribution, then the PEM can be used as a means of fitting the distribution. For example, suppose we suspect that the failure function obeys a Gumbel distribution

$$F_G(g) = e^{-e^{-\frac{(g-a)}{b}}} \tag{5.54}$$

This has mean μ_G and variance σ_G^2, given by

$$\mu_G = a - b\Gamma'(1) \tag{5.55}$$
$$\sigma_G^2 = (b\pi)^2/6 \tag{5.56}$$

where $\Gamma'(x)$ is the first derivative of the gamma function (see e.g., Abramowitz & Stegun 1964), and $\Gamma'(1) = -0.57721$.

The PEM gives estimates of μ_G and the second moment of G, from which the variance σ_G^2 may be computed (see Section 2.3.3). Substituting these estimates into the left-hand sides of Equations (5.55) and (5.56) gives two simultaneous equations that may be solved to obtain the two unknown parameters a and b. The probability of failure is then estimated from $P(G < 0) = F_G(0)$.

5.7 Level 3 methods

Level 3 methods are the most general of the reliability techniques. The approach in Level 3 methods is to obtain an estimate of the integral in Equation (5.4) through numerical means. The complexity of the integral (in general) means that numerical, rather than analytical, methods are used. There are two widely used techniques:

1 Monte Carlo integration;
2 Monte Carlo simulation.

The first method may be used if there is a closed analytical form for the probability distribution of the reliability function and a failure region that is well defined in terms of the basic variables. As illustrated in Figure 5.13 for the two-dimensional case, Monte Carlo integration evaluates the function at a random sample of points, and estimates its integral based on that random sample (Hammersley & Handscomb 1964, Press et al. 2007). In symbols:

$$\int_{G<0} f_G(\underline{x})\,d\underline{x} \approx V <f> \pm V\sqrt{\frac{<f^2>-<f>^2}{N}} \tag{5.57}$$

where

$$<f> \equiv \frac{1}{N}\sum_{i=1}^{N} f(x_i) <f^2> \equiv \frac{1}{N}\sum_{i=1}^{N} f^2(x_i) \tag{5.58}$$

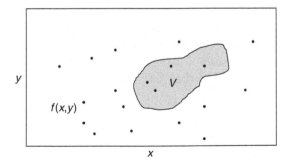

Figure 5.13 Illustration of Monte Carlo integration. The function $f(x,y)$ is evaluated at randomly spaced points in a 'volume' defined in $x-y$ space. Only those points lying within the volume V for which $G<0$ are used. The choice of volume in this case could be considerably reduced to improve the efficiency of the process.

and x_i, for $i = 1, \ldots, N$, are random points uniformly distributed in the multidimensional volume V for which $G < 0$.

Example 5.4. Estimate the integral of the function $y = f(x) = 4x^3 - 3x^2 + 1$ over the region $x \in [0, 2]$; that is,

$$\int_0^2 (4x^3 - 3x^2 + 1) dx$$

exactly and also using Monte Carlo integration using 2, 4 and 8 points from the random number generation example (Example 2.17).

Solution. The exact value of the integral may be found by direct integration and is 10. We next perform Monte Carlo integration using the first 2, first 4 and all 8 of the random numbers in Table 2.2. The 'volume' V in this case is simply the length of the line segment $[0, 2]$, that is, 2.

For $N = 2$: First, we use the first two random numbers in Table 2.2 (0.75 and 0.875). Now, these were generated for a random variate lying between 0 and 1. To cover the interval $[0, 2]$, we have to multiply all the random numbers by 2. Thus, we have $f(1.5) = 7.75$ and $f(1.75) = 13.25$. From Equations (5.57 and 5.58) we have:

$$\int_0^2 f(x)\, dx \approx V < f > \pm V \sqrt{\frac{<f^2> - <f>^2}{N}} = 2 \left(\frac{1}{2} \{7.75 + 13.25\} \right)$$

$$\pm 2 \sqrt{\frac{(7.75^2 + 13.25^2)/2 - ((7.75 + 13.25)/2)^2}{2}}$$

$$= 21.0 \pm 15.8$$

The error bounds are very large and should alert us to the very approximate nature of the result. For $N = 4$: We take the first four values in Table 2.2 and repeat the process used for the case $N = 2$. Thus, we already have $f(1.5) = 7.75$ and $f(1.75) = 13.25$, and we find that $f(1) = 1.0$ and $f(1.25) = 4.125$. Hence,

$$\int_0^2 f(x)\, dx \approx V < f > \pm V \sqrt{\frac{<f^2> - <f>^2}{N}}$$

$$= 2 \left(\frac{1}{4} \{7.75 + 13.25 + 1.0 + 4.125\} \right) \pm 2 \sqrt{\frac{63.46 - (6.54)^2}{4}}$$

$$= 13.06 \pm 4.56$$

Finally, for $N = 8$, we find:

$$\int_0^2 f(x)\,dx \approx V <f> \pm V\sqrt{\frac{<f^2> - <f>^2}{N}}$$

$$= 2\left(\frac{1}{8}\{7.75 + 13.25 + 1.0 + 4.125 + 0.75 + 1 + 1 + .875\}\right)$$

$$\pm 2\sqrt{\frac{32.12 - (3.72)^2}{8}}$$

$$= 7.44 \pm 3.02$$

It is worth noting a couple of points. First, in this example, the error bounds more than cover the discrepancy between the estimated integral and the exact solution. Second, the estimates can be either greater or smaller than the exact result.

The 'error bounds' defined in this way are simply an estimate of one standard deviation either side of the mean and, critically, depend upon both the x_i and N. Clearly, the larger N the better the estimate of the integral should be. However, as the number of dimensions increases, the number of points required goes up as the power of the dimension. The consequent computational effort can be mitigated somewhat by a technique known as *importance sampling*. In essence, this involves selecting the random points so that, rather than being uniformly distributed, they follow a distribution designed to make the integrand appear as constant as possible. (See also Further Reading at the end of this chapter.)

In Monte Carlo simulation a set of values of the basic variables is generated with the appropriate probability distribution and values of the reliability function determined. By repeating this process many times and storing the results, the integral may be estimated as the proportion of the results for which the reliability function is negative. In symbols, if X_n is the nth simulation, then the Monte Carlo estimate of the integral is

$$\frac{\text{number of the } X_n \,(n = 1, \ldots, N) \text{ in the failure region}}{\text{total number of simulations } (= N)} \quad (5.59)$$

Clearly, increasing N improves the precision of the answer and, in practice, N should correspond to at least 10 times the length of the return period of interest.

Evidently, large sample sizes are required for the most extreme events, which can be computationally demanding. There are methods available, known as *variance reduction techniques*, for improving precision without increasing N. These methods use a disproportionate number of extreme conditions at the simulation stage, but the manner in which this is done is not straightforward.

5.8   Further notes on reliability methods

In this section two additional topics are discussed: second-order reliability methods and time-varying reliability methods.

5.8.1   *Second-order reliability methods*

In the previous sections, the focus has been on first-order reliability methods (FORMs). These are based on linearising the nonlinear reliability function about the point of interest via truncation of the Taylor series expansion. This is equivalent to finding the tangent line or plane to the reliability surface. A natural progression is to retain additional terms in the Taylor series expansion. Methods that include nonlinear approximations to the reliability function have been termed second-order reliability methods (SORMs) by Fiessler et al. (1979) and Hohenbichler et al. (1987). By retaining an additional term in the Taylor series expansion, a quadratic approximation to the reliability surface may be obtained. The advantage of this is that the sense of curvature of the reliability surface can be captured. Whether this is important or not will clearly depend on the nature of the reliability surface in the region of the design point. The situation is illustrated in Figure 5.14. In the case of a linear approximation (Figure 5.14a), errors locating the exact position of the design point can result in an over- or underestimate of the probability of failure, depending on whether the reliability surface is convex or concave, respectively. This can be mitigated to some extent by means of a higher order approximation (Figure 5.14b), which captures the sense of the curvature in the surface.

Of course, a quadratic approximation to the surface cannot capture any skewness, for which a third-order approximation is required, and so on. The quantity of algebra involved, and the often modest improvements in the estimates of probability of failure, mean that retaining extra terms follows a law

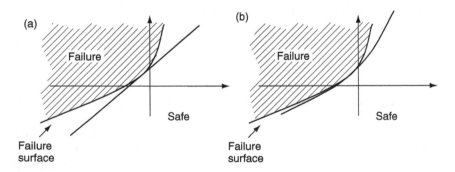

Figure 5.14 (a) Linear approximation to the reliability surface at the design point; (b) quadratic approximation to the reliability surface at the design point.

of rapidly diminishing returns. Alternative approaches to obtaining higher-order estimates of the probability of failure can be found in the literature (e.g., Breitung 1984, Abdo & Rackwitz 1990, Breitung 1994, Ditlevsen & Madsen 1996).

A note of pragmatism: the accuracy of any estimate of the probability of failure will be conditioned by many other factors, including: the quality of the data; the methods used to analyse the data (say, in fitting probability distribution functions); the quality of the design formulae used in formulating the reliability function; and the quality of materials and construction. The level of probabilistic analysis in a reliability calculation should be chosen in the context of the problem at hand, that is, the method should be commensurate with the quality and accuracy of the data and design formulae. For example, if the design formula used in the reliability function is known to be accurate only to within a factor of 50%, it is hardly appropriate to expect estimates of the probability of failure to have better accuracy than this, and the use of high-order reliability methods is unlikely to be cost effective.

5.8.2 Time-varying reliability methods

The alert reader will have noticed that in the foregoing discussion time has been an implicit rather than explicit variable. From experience it is known that not all failures result from spectacular extreme loading events, but rather occur gradually over time, either as the integrity of a structure degrades due to wear and tear, or as loading conditions gradually increase, perhaps as a result of changes in climate. To describe the basic reliability problem (Equation 5.3) in these terms requires the strength and loading functions to be considered as a function of time:

$$P_f(t) = P[R(t) < S(t)] = P[G(t) < 0] \tag{5.60}$$

so that the strength, the load, their probability distribution functions and the probability of failure are all functions of time. Figure 5.15 illustrates this scenario.

Although $R(t)$ and $S(t)$ may vary continuously in time, from a practical point of view any time series will actually have a finite resolution determined by the resolving power of a measuring instrument. The instantaneous probability of failure may also be written in terms of the density function of the reliability function:

$$P_f(t) = \int_{G(X_1,X_2,\ldots,X_n)<0} f_{X_1,X_2,\ldots,X_n}[x_1(t),x_2(t),\ldots,x_n(t)]$$
$$dX_1(t)dX_2(t)\ldots dX_n(t) \tag{5.61}$$

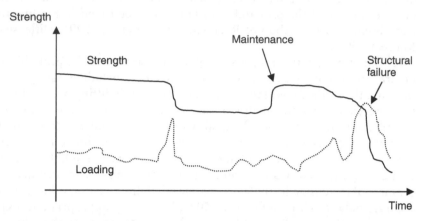

Figure 5.15 Illustration of the time-varying nature of loads and structural strength.

where each of the basic variables $X_1, X_2, \ldots, X_n$ is considered to be a function of time. The probability of a failure occurring over a period of time, say $t = 0$ to $t = T$, is given by integrating over this period with respect to time:

$$P_f(T) = \int_{t=0}^{t=T} \int_{G(X_1,X_2,\ldots,X_n)<0} f_{X_1,X_2,\ldots,X_n}[x_1(t), x_2(t), \ldots, x_n(t)]$$

$$dX_1(t)dX_2(t)\ldots dX_n(t)dt \tag{5.62}$$

This is a very general statement of the reliability problem, involving the integration of a function of many potentially correlated (both in time and with each other) stochastic random variables.

A common simplifying approach is to consider time-integrated quantities, that is, attention is focused on the whole lifetime of the structure, $t \in [0, t_L]$, and all random variables relate to this lifetime. Thus, the probability of interest with regard to loads is the maximum load over the lifetime, and that for strength is the minimum strength over the lifetime. In many cases the strength is considered time invariant, and the load can be represented over the total time period by an extreme value distribution.

A common variant of this approach that allows some time variation is to integrate over 'blocks' of time. The lifetime of the structure is then composed of a succession of blocks. A typical length for a block is a year, and the probability of failure in each block refers to the probability that the annual maximum of the load exceeds the strength. A further assumption is that the annual maxima of the loads are identically distributed and independent. Each year may be viewed as 'an event' in the parlance of discrete variables (see Chapter 2). Thus, the probability that the maximum load is less than some value, say a, for any year is given by the cumulative distribution function $F_S(S < a)$.

Example 5.5. If the annual probability of a load exceeding a critical value a is p, and the design life of the structure is t_L, what is the probability of the load exceeding the critical value during the lifetime of the structure?

Solution. The probability of exceedance over t_L years is (1 – the probability of non-exceedance over t_L years). The probability of non-exceedance in any year is $1 - p$. As each year is considered to be independent, the probability of non-exceedance in n years is $(1 - p)^n$. Thus, the probability of the load exceeding the critical value during t_L years is $1 - (1 - p)^{t_L}$. (See also Section 4.1.)

Example 5.6. Taking the situation described in Example 5.5, let $p = 0.01$, which corresponds to a 1 in 100 year event, and let the lifetime of the structure be 50 years.

a Calculate the probability that the critical value is exceeded in 10, 50 and 100 years.
b Calculate the probability that the critical value is exceeded in the first 2 years only of the lifetime of the structure.
c Calculate the probability that the critical value is exceeded four times during the lifetime of the structure.

Solution.

a Following the argument in the previous example, the probabilities are given by $1 - (1 - p)^n$, where $n = 10$, 50 and 100. Evaluating the expression for these values of n gives 0.0956, 0.395 and 0.634, respectively, illustrating that the longer the period under consideration, the greater the probability of excess.
b The required probability is given by the product of the probability of exceedance in each of the first 2 years and non-exceedance in each of the following 48 years, that is, $p.p.(1 - p)^{48} = 6.17 \times 10^{-5}$
c This is a combination-type problem (see Chapter 2). The probability is given by

$$^nC^r.p^4(1 - p)^{46} = {}^{50}C^4 \times 6.30 \times 10^{-9} = 230,300 \times 6.30 \times 10^{-9}$$
$$= 1.45 \times 10^{-3}$$

which is the probability of obtaining four exceedances in 50 'events' in every possible combination.

It may seem surprising that the probability in (b) is so much smaller than that in (c). The reason for this is that, in (b), the probability refers to exceedances occurring in two specific years, while in (c) the four exceedances can occur in any years during the lifetime of the structure.

An alternative to the 'block' approach is to consider individual load events that are likely to take the structure close to the ultimate limit state. For flooding problems this will normally equate to major storms. A common assumption is to consider storms to be independent, non-overlapping and randomly occurring. A useful model for this type of behaviour is the Poisson distribution (see Chapter 2). The probability, $p_k(t)$, of k events occurring in the interval $[0, t]$ is defined as:

$$p_k(t) = \frac{(at)^k e^{-at}}{k!} \tag{5.63}$$

and is illustrated diagrammatically in Figure 5.16.

The probability of failure over the period $[0, t_L]$ may then be determined as

$$P_f(t) = \int_0^\infty \sum_{k=0}^\infty p_k(t)[1 - F_S(y)]f_R(y)dy \tag{5.64}$$

where y is the basic variable, the summation is over all possible numbers of storms that can occur in the interval, and the probability functions are understood to be extreme distributions.

A further alternative approach is to treat the loads, strengths and thus the reliability function as time-dependent stochastic processes. Failure is defined as the point at which the reliability function first crosses from being positive to negative. This is sometimes referred to as the 'first passage' or 'outcrossing' problem in stochastic process theory. Details of the theory are beyond the scope of this book, but a brief introduction to stochastic processes was given in Chapter 3. Typically, the theory will yield the mean and variance of the time to the first outcrossing, that is, the expected time to failure, together with a measure of the dispersion. The theory is far from straightforward, even for relatively restrictive cases. For practical hydraulics problems, where the basic variables may be correlated, nonstationary and non-normal, a practical alternative is Monte Carlo simulation, but care must be taken to

Figure 5.16 Example of occurrences of a Poisson process in time.

ensure that the statistical properties of the randomly generated sequences of variables actually reflect the desired properties. A means of generating normally distributed variables with arbitrary temporal autocorrelation can be found in Chapter 3. In general, variables will not be normally distributed, and a means of generating correlated non-normal multivariate sequences is required, such as that proposed by Cai et al. (2008). In some cases, it may be possible to obtain an analytical or semi-analytical solution to a time-varying problem. An example of how such solutions can be used in a reliability problem is given in Chapter 6, in the section on coastal morphology (Section 6.3.1).

This section concludes with analytical results for Poisson and geometrically distributed processes. The cumulative distribution function $F_W(t)$ for the time W_n that elapses before the occurrence of the nth event in a Poisson process is given by

$$
\begin{aligned}
F_W(t) &= 1 - P(W_n > t) \\
&= 1 - P(\text{number of events} < n) \\
&= 1 - \sum_{k=0}^{n-1} \frac{(at)^k e^{-at}}{k!} \\
&\equiv P(k, at) \qquad t \geq 0, k > 0 \\
&= \frac{1}{\Gamma(k)} \int_0^{at} e^{-s} s^{k-1} ds
\end{aligned}
\tag{5.65}
$$

where $\Gamma(k)$ is the gamma function, $P(k, at)$ is the incomplete gamma function, and $F_W(t)$ is the Gamma distribution with mean n/a and variance n/a^2. F_W is the distribution function for the Gamma density function (Equation 3.61). For the case where $n = 1$, this simplifies to

$$
F_W(t) = 1 - e^{-at}.
$$

In a sequence of Bernoulli trials (see Chapter 2), the probability that an event will occur for the first time during trial N is given by the geometric distribution (see Equation 2.14) as:

$$
P(\text{First occurrence is for } n = N) = p(1 - p)^{N-1}
$$

- This can be used to estimate the *average* time between events, that is, the return period. Now,

$$\text{Expected time between events} = E(T) = \sum_{n=1}^{\infty} np(1-p)^{n-1}$$

$$= p[1 + 2(1-p) + 3(1-p)^2 + \ldots]$$

$$= p\left[\frac{1}{(1-(1-p))^2}\right]$$

$$= p\left[\frac{1}{p^2}\right]$$

$$= \frac{1}{p}$$

where we have used a series expansion result in going from line 2 to line 3. A flood level with an annual exceedance probability of p will be exceeded at an *average* rate of once every $1/p$ years. A typical benchmark for catastrophic flooding is the 1 in 100 year event. This is a severity of flood that has a probability of being exceeded in any given year of 0.01. It is important to note that several 100-year floods can occur within a few years, and conversely several hundred years may pass without a 1 in 100 year flood occurring. The long-term average rate will be once every 100 years.

Further reading

Lepage, G. P., 1978. A new algorithm for adaptive multidimensional integration. *Journal of Computational Physics*, 27: 192–203.

Liu, J. S., 2002. *Monte Carlo strategies in Scientific Computing*, Springer, New York.

Melchers, R. E., 1999. *Structural Reliability Analysis and Prediction*, 2nd edition, John Wiley & Sons, Chichester, p. 437.

Nowak, A. S. and Collins, K. R., 2000. *Reliability of Structures*, McGraw-Hill, Boston.

Thoft-Christensen, P. and Baker, M. J., 1982. *Structural Reliability Theory and Its Applications*, Springer, Berlin.

Wen, Y. K. and Chen, H. C., 1987. On fast integration for time variant structural reliability. *Probabilistic Engineering Mechanics*, 2: 156–162.

6 Applications

6.1 Introduction

This chapter is organised as a set of worked examples. Each section covers problems in a particular discipline; but the problems cover a range of types of reliability analysis. The majority of these have been described in detail in the previous chapters; where they have not, additional working is provided. It is anticipated that the reader will not necessarily be familiar with the appropriate background in all disciplines covered in this chapter; hence each section includes the relevant formulae, equations and background concepts used in the examples. References to additional sources of material are provided at the end of the chapter for those readers seeking further information in particular areas. Each example is structured in the following manner. First, a statement and explanation of the problem is given, including a discussion of possible failure modes and the particular mode(s) considered. Then, the failure criterion is defined, together with the reliability function, and the basic variables of the problem. Next, the method of solution is identified (e.g., Level 1, Level 2), and, finally, the worked solution is presented.

6.2 Fluvial flood defences

6.2.1 Introduction

The reliability of fluvial flood defences is governed mainly by the water level and the flow speed. Flooding occurs when the water level exceeds the crest level of an embankment, while excessive flow speed can lead to scouring around the toe of a structure, destabilising it. Additional complications can arise from flow of water through or under the materials comprising the embankment. Here, we restrict attention to overflow and scour.

The Froude number F_r is defined as

$$F_r = V/(gR)^{1/2} \tag{6.1}$$

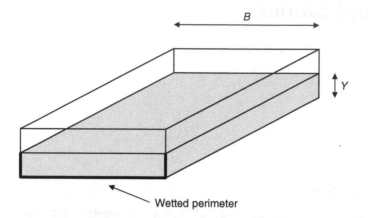

Figure 6.1 Definition of variables for open channel flow.

where V is the flow velocity, g the acceleration due to gravity $(9.81\,\text{ms}^{-2})$ and R is the hydraulic radius defined as the ratio of the cross-sectional area of flow to the wetted perimeter. The hydraulic radius is an important parameter in hydraulic engineering.

For a rectangular channel of width B and water depth $Y, R = BY/(B + 2Y)$, see Figure 6.1. For a wide channel $B >> Y$ and $R \approx Y$. Manning's formula for the velocity in an open-channel cross-section is

$$V = \frac{S^{1/2}R^{2/3}}{n} \qquad (6.2\text{a})$$

where n is the roughness coefficient and S is the friction slope of the channel. For a wide rectangular channel this is equivalent to

$$V = \frac{S^{1/2}Y^{2/3}}{n} \qquad (6.2\text{b})$$

The alert reader will have noticed that the symbols R and S, defined as strength and load in the discussion of reliability theory in Chapter 5, are now being used for different quantities, in accordance with common usage in hydraulics. This is one of the difficulties when trying to combine two disciplines that have developed separately. Here, we use the symbols for multiple quantities, clarifying the usage in particular instances.

When water flows around objects, such as bridge piers, the nature of the flow can change. Thus, rather than being smooth (or laminar), the flow may become confused and chaotic (or turbulent). This can lead to grains of sediment being moved by the flow, and eventually to scouring. Figure 6.2 illustrates the flow around a bridge pier and the corresponding pattern of scour.

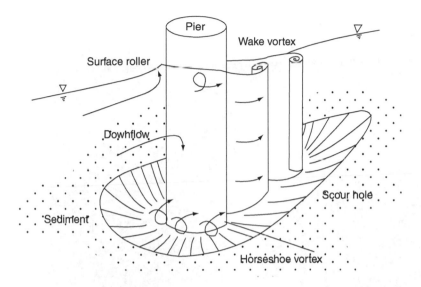

Figure 6.2 Diagrammatic representation of the flow pattern and scour hole at a cylindrical bridge pier. The principal causes of scour are the horseshoe and wake vortices, together with the downflow.

The empirical formula for estimating scour depth near piers, developed by Johnson (1992), is

$$D = 2.02\,Y(b/Y)^{0.98}\,F_r^{\,0.21}\,W^{-0.24} \tag{6.3}$$

where the scour depth is D, b is the pier width, Y is the flow depth immediately upstream of the pier, W is a measure of the sediment size ($W = D_{84}/D_{50}$, the ratio of the 84% quantile to the median sediment diameter). The constant of proportionality (2.02) is appropriate for quantities measured using metric units (metres).

6.2.2 *Examples*

Example 6.1. Overflow of an embankment

Statement of problem – An earth embankment for a river reach is required to prevent flooding such as shown in Figure 6.3.

Observations have shown that water levels in the river are usually below +4.5 m relative to a local datum. Recent measurements have shown that town development in upstream reaches has led to a 'peakier' hydrograph, and it has been recommended by the local river authority that a factor of safety of 1.07 should be used to account for the higher water levels associated with these peaks. Experience in designing and constructing

Figure 6.3 Overflow of Robinson's Marsh embankment, Suffolk, UK. Erosion of the backface of the embankment is evident and is a precursor of the formation of a breach (courtesy Halcrow Group Ltd).

earth embankments has shown that account must be taken of: variations in construction accuracy; variations in soil type; settlement; burrowing by rabbits, which can lead to local collapse of the embankment and reduction in crest level; and compaction by pedestrians and vehicles moving along the top of the embankment. The corresponding recommended partial safety factors are: 1.05, 1.1, 1.12, 1.05 and 1.20, respectively. Determine the global factor of safety and thus the crest level required.

The *failure criterion* is thus that the water level exceeds the crest level. The *basic variables* are water level and embankment crest level. This is a problem concerning *Level 1 methods* that require specification of safety factors.

Solution. Using the nomenclature of reliability theory, let the crest level be R and the water level be S. Then, from Equation (5.6) we have:

$$\frac{R}{\Gamma_r} = \Gamma_{s1}\, \Gamma_{s2}\, \Gamma_{s3}\, \Gamma_{s4}\, \Gamma_{s5} S \qquad (6.4)$$

where Γ_r is the factor of safety for the water levels and Γ_{si} are the factors of safety for the embankment. The global factor of safety is given by the product of all the factors of safety $= 1.07 \times 1.05 \times 1.1 \times 1.12 \times 1.05 \times 1.20 = 1.74$. Substituting in the values of the factors of safety, together with the value of 4.5 as the nominal value of the water level, gives the required crest level as:

$$R = \Gamma_r \, \Gamma_{s1} \, \Gamma_{s2} \, \Gamma_{s3} \, \Gamma_{s4} \, \Gamma_{s5} \, S = 1.07 \times 1.05 \times 1.1 \times 1.12 \times 1.05 \times 1.20$$
$$\times 4.5 = 7.85 \, \text{m}$$

This is significantly higher than the nominal water level. The more factors that are included, the greater the global factor of safety, because each individual partial safety factor is greater than unity. When using this type of approach, care and judgement are necessary, both in the identification of variables and in the definition of partial safety factors, in order to obtain a solution that is both safe and economic.

Example 6.2. Overflow of an embankment

Statement of problem – Consider the same problem as described in Example 5.1, with an alteration. Namely, the crest level of embankment overreach is described deterministically as 5 m rather than by a normal distribution with a mean of 5 m and a standard deviation of 0.5 m. Monthly maximum water levels along the reach obey $N(3, 1)$. What is the probability of flooding?

The *failure criterion* is, as in the previous example, that the water level exceeds the crest level. The *basic variables*, water level and crest level, are described by probability density functions, therefore a Level 2 method will be appropriate.

Solution. Flooding occurs when water level > crest level. Using the nomenclature of reliability theory, let the crest level be R and the water level be S. So the reliability function can be written as

$$G = \bar{R} - S \tag{6.5}$$

with failure occurring when $G < 0$. Two approaches are possible. Either treat the crest level as a $N(5, 0)$ distribution (i.e., it has mean of 5 and zero variance), or note that the problem reduces to finding the probability that the water level exceeds 5 m. In the first case, the variables are normal and independent, so (from Equations 5.7 and 5.8)

$$\mu_G = 5 - 3 = 2, \ \sigma_G^2 = 0^2 + 1^2 = 1 \ \Rightarrow \beta = 2 \tag{6.6}$$

And so, from Equation (5.9), we have that the probability of failure is given by:

$$P_F = \Phi\left(\frac{0-2}{1}\right) = \Phi(-2) = 0.023 \approx 2\% \tag{6.7}$$

Alternatively, note that failure occurs when the water level exceeds 5 m, and thus the probability of failure is just the probability that the water level is greater than 5 m. This can be computed from the tables in Appendix A by first standardising the water level variable by setting $z = (S - \mu_S)/\sigma_S = (S - 3)/1$. The probability of failure is then the probability that $S > 5$, or that $z > 2$. In other words, from Table D in Appendix A, the probability of failure is 0.023.

Thus, the *probability of failure* is approximately 2% *per month*. This is about half the probability of failure of the structure in Example 5.1. The reason why the probability is smaller in this case is because the variance of the reliability function G is less while the mean is the same. There is thus less area under the probability density curve to the left of the y-axis (viz. Figure 5.6).

Example 6.3. Scour around bridge piers

Statement of problem – Local scour around bridge piers can undermine the foundations and lead to collapse (Figure 6.4). Given the following hydraulic and sediment information, estimate the mean and variance of the scour depth. The pier width is 3.0 m, the channel slope is $S \sim N(0.002, 0.0004)$, the depth Y has exponential density with parameter 0.2, the roughness coefficient has a uniform distribution between 0.02 and 0.05, and the sediment grading is $W \sim N(4, 2)$. It may also be assumed that S, Y, b and n are independent. The *failure criterion* is not defined in this case, but would be a depth of scour considered to compromise the integrity of the bridge pier foundation. To estimate the scour depth we use Johnson's formula (Equation 6.3) and Manning's formula (Equation 6.2) for the velocity. The *basic variables* are those required to evaluate Johnson's formula.

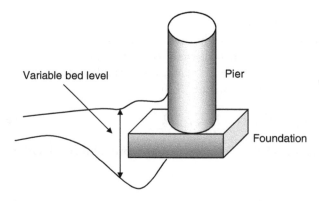

Figure 6.4 Illustrating uncertainty in scour depth.

The *method of solution* will be to use Equation (3.64) to estimate the mean and variance of the scour from the partial derivatives. This is a *Level 2 type approach*, and will yield an approximate solution.

Solution. Substituting Manning's formula into the expression for the Froude number (Equation 6.1) gives

$$F_r = S^{1/2} Y^{1/6} n^{-1} g^{-1/2}$$

Thus, the scour depth, D, in Johnson's formula, can be written as

$$D = 1.59 b^{0.98} Y^{0.055} S^{0.105} n^{-0.21} W^{-0.24} \tag{6.8}$$

From Equation (3.56) note that the mean and standard deviation of Y are 5 m and $\sqrt{5}$ m, respectively. The mean and standard deviation of the roughness coefficient are 0.03 and 0.0087 (see Equation 2.30). To use Equation (3.64) to calculate estimates of the mean and variance of D requires the first and second derivatives, evaluated at mean values:

$$\left.\frac{\partial D}{\partial y}\right|_{\mu} = 1.59 \times 0.055 b^{0.98} \mu_Y^{-0.945} \mu_S^{0.105} \mu_n^{-0.21} \mu_W^{-0.24} = 0.04245$$

$$\left.\frac{\partial D}{\partial s}\right|_{\mu} = 1.59 \times 0.105 b^{0.98} \mu_Y^{0.055} \mu_S^{-0.895} \mu_n^{-0.21} \mu_W^{-0.24} = 202.16$$

$$\left.\frac{\partial D}{\partial n}\right|_{\mu} = -1.59 \times 0.21 b^{0.98} \mu_Y^{0.055} \mu_S^{0.105} \mu_n^{-1.21} \mu_W^{-0.24} = -23.11$$

$$\left.\frac{\partial D}{\partial w}\right|_{\mu} = -1.59 \times 0.24 b^{0.98} \mu_Y^{0.055} \mu_S^{0.105} \mu_n^{-0.21} \mu_W^{-1.24} = -0.2311$$

$$\left.\frac{\partial^2 D}{\partial y^2}\right|_{\mu} = -1.59 x 0.052 b^{0.98} \mu_Y^{-1.945} \mu_S^{0.105} \mu_n^{-0.21} \mu_W^{-0.24} = 0.00801$$

$$\left.\frac{\partial^2 D}{\partial s^2}\right|_{\mu} = -1.59 \times 0.094 b^{0.98} \mu_Y^{0.055} \mu_S^{-1.895} \mu_n^{-0.21} \mu_W^{-0.24} = -90477.5$$

$$\left.\frac{\partial^2 D}{\partial n^2}\right|_{\mu} = 1.59 \times 0.254 b^{0.98} \mu_Y^{0.055} \mu_S^{0.105} \mu_n^{-2.21} \mu_W^{-0.24} = 798.68$$

$$\left.\frac{\partial^2 D}{\partial w^2}\right|_{\mu} = 1.59 \times 0.298 b^{0.98} \mu_Y^{0.055} \mu_S^{0.105} \mu_n^{-0.21} \mu_W^{-2.24} = 0.07174$$

where μ_Y, μ_S, μ_n and μ_W are the mean of Y, S, n and W, respectively. The first-order estimate of the mean scour depth is obtained by substituting the mean values of the variables into Equation (6.8):

$$E(D) \approx 3.85$$

The second-order estimates of the mean and variance are given by

$$E(D) \approx 3.85 + 0.5(0.00801 \times 5 - 90477.5 \times 0.0004^2 + 798.68$$
$$\times 0.0087^2 + 0.07174 \times 2^2) = 4.036\,\text{m}$$
$$\text{Var}(D) \approx 0.04245^2 \times 5 + 202.16^2 \times 0.0004^2 + (-23.11)^2 \times 0.0087^2$$
$$+ (-0.2311)^2 \times 2^2 = 0.270\,\text{m}^2$$

Example 6.4. Overflow of flood embankment

Statement of problem – In the early stages of the development of reliability theory, the 'factor of safety' was often used to define the reliability of a structure (see Section 5.5). The factor of safety was defined as the ratio of the assumed nominal values of strength and loading. Considering the strength and loading to be random variables R and S, respectively, the factor of safety may be written as $Z = R/S$. R and S are described by probability distributions and, through the techniques described in Chapter 3, the distribution of Z may also be found. The probability of failure is then given by the probability that the loading is greater than the strength, or $Z < 1$. Consider an embankment whose crest level is R and water levels are S, where R and S are independent log-normal variables, with means and standard deviations of $\mu_R = 9.0\,\text{m}, \mu_S = 6.7\,\text{m}, \sigma_R = 1.0\,\text{m}$ and $\sigma_S = 1.5$ m, respectively. The *failure criterion* is thus that the water level exceeds the crest level, and these two variables are the *basic variables* of the problem. Find the probability that the water level exceeds the embankment crest. We are given information about the probability distributions of the basic variables, so a *Level 2 approach* is an appropriate method of solution.

Solution. We write $Z = R/S$, and determine the probability that $Z < 1$.

It is left as an exercise for the reader to show that, if R and S are lognormal, then so is Z. Now, $\ln(Z) = \ln(R) - \ln(S)$, so (from Equation 3.59)

$$\mu_{\ln(Z)} = \mu_{\ln(R)} - \mu_{\ln(S)} = \ln(\mu_R) - 0.5\ln\left[1 + (\sigma_R/\mu_R)^2\right] - \ln(\mu_S)$$
$$+ 0.5\ln\left[1 + (\sigma_S/\mu_S)^2\right]$$

Similarly,

$$\sigma_{\ln(Z)}^2 = \sigma_{\ln(R)}^2 + \sigma_{\ln(S)}^2 = \ln\left[1 + (\sigma_R/\mu_R)^2\right] + \ln\left[1 + (\sigma_S/\mu_S)^2\right]$$

Since $\ln(Z)$ is normally distributed, with mean $\mu_{\ln(Z)}$ and standard deviation $\sigma_{\ln(Z)}$, the random variable $[\ln(Z) - \mu_{\ln(Z)}]/\sigma_{\ln(Z)}$ is a standard normal variable. Therefore, the probability of failure may be found as:

$$P_f = F_Z(1) = \Phi\left(\frac{\ln(1) - \mu_{\ln(z)}}{\sigma_{\ln(Z)}}\right) = \Phi\left(-\frac{\mu_{\ln(z)}}{\sigma_{\ln(Z)}}\right)$$

Substituting the numerical values in for the mean and standard deviations yields

$$\mu_{\ln(Z)} = \ln(9.0) - 0.5 \ln\left[1 + (1.0/9.0)^2\right] - \ln(6.7)$$
$$+ 0.5 \ln\left[1 + (1.5/6.7)^2\right]$$
$$= 2.197 - 0.00615 - 1.902 + 0.0245$$
$$= 0.3134$$

$$\sigma^2_{\ln(Z)} = \ln\left[1 + (1.0/9.0)^2\right] + \ln\left[1 + (1.5/6.7)^2\right]$$
$$= 0.0123 + 0.0489 = 0.0612$$

The probability of failure is thus $P_f = \Phi(-0.3134/0.2474) = \Phi(-1.267) = 1 - \Phi(1.267) \approx 1 - 0.898 \approx 0.1$.

Example 6.5. Irrigation scheme

Statement of problem – An irrigation scheme has a demand Y of water from a river (Figure 6.5). The mean demand is 15 units, with a standard deviation of 5 units, which accounts for variations in water supply due to rainfall variability. The mean quantity of water available for abstraction, X, is 20 units, with a standard deviation of 3 units, which accounts for the seasonal hydrological variability. Due to the link between hydrology and climate, the

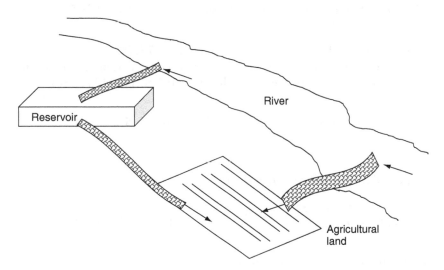

Figure 6.5 Illustrating abstraction from a river plus an additional reservoir to smooth out peaks in demand and supply.

natural water availability often tends to decrease when demand increases, so that the random variables X and Y are negatively correlated, with an estimated correlation coefficient of -0.6. If both X and Y are normally distributed, what is the reliability of the system? If, subsequently, a reservoir is constructed that weakens the correlation between supply and demand to -0.2, what effect does this have on the reliability? What is the result of ignoring the correlation?

The *failure criterion* is that the water demand exceeds the water supply. The *basic variables* are water demand and supply. Information about the distribution and correlation between the basic variables is provided, so a *Level 2 type method* of solution is appropriate.

Solution.   The reliability function may be written as $G = X - Y$. As X and Y are both normally distributed, so is G (see Examples 3.9 and 3.13). Furthermore,

$$\mu_G = \mu_X - \mu_Y = 20 - 15 = 5$$

and

$$\sigma_G = \left(\sigma_X^2 - 2\rho_{XY}\sigma_X\sigma_Y + \sigma_Y^2\right)^{1/2} = (25 - 2 \times (-0.6) \times 5 \times 3 + 9)^{1/2}$$
$$= 7.21\,\text{units}$$

The probability of failure, P_f, is

$$P_f = F_G(0) = \Phi\left(-\frac{5}{7.21}\right) = 1 - \Phi\,(0.693) \approx 1 - 0.755 = 0.245$$

The reliability of the system is $1 - P_f = 0.755$, or 75.5%.

If the correlation is then altered to -0.2 by the construction of a reservoir, the mean values remain the same, but the standard deviation becomes

$$\sigma_G = \left(\sigma_X^2 - 2\rho_{XY}\sigma_X\sigma_Y + \sigma_Y^2\right)^{1/2} = (25 - 2 \times (-0.2) \times 5 \times 3 + 9)^{1/2}$$
$$= 6.08\,\text{units}$$

The probability of failure, P_f, is then

$$P_f = F_G(0) = \Phi\left(-\frac{5}{6.08}\right) = 1 - \Phi\,(0.822) \approx 1 - 0.794 = 0.206$$

The reliability of the system is $1 - P_f = 0.794$, or 79.4%. Thus, the construction of a reservoir to alter the correlation properties does not have a very large impact in this case. Should the correlation between the variables be ignored, that is, by setting $\rho_{XY} = 0$, then the standard deviation of G is given by:

$$\sigma_G = \left(\sigma_X^2 + \sigma_Y^2\right)^{1/2} = (25 + 9)^{1/2} = 5.83 \text{ units}$$

The probability of failure P_f is then estimated as

$$P_f = F_G(0) = \Phi\left(-\frac{5}{5.83}\right) = 1 - \Phi(0.858) \approx 1 - 0.805 = 0.195$$

The reliability of the system is $1 - P_f = 0.805$, or 80.5%. Thus, in this case, ignoring the correlation leads to an *overestimate* of the reliability of the scheme.

Example 6.6. Water quality

Statement of problem – Water quality is important for both bathing water as well as drinking water. Criteria for safe water have been defined by legislation in many continents, including Europe and America (e.g., Council Directives 1975, 1991, EPA 2007). This example provides an illustration of how reliability methods can be applied to such issues. For the sake of argument, the quantity of phytoplankton (algae) present in the water is used as the critical variable. The quantity of phytoplankton will depend on the availability of nutrients, temperature, daylight, and depredation due to death and grazing by zooplankton. Here, a simple model for the growth of phytoplankton is used, which is a linear function of nutrient concentration X_1, temperature X_2 and solar radiation X_3, and depredation as a constant value a_0, with units of mg/m³. We take $X_1 \sim N(100 \text{ mg/m}^3,\ 75 \text{ mg/m}^3)$, $X_2 \sim N(16°C,\ 6°C)$ and $X_3 \sim N(130 \text{ W/m}^2,\ 70 \text{ W/m}^2)$. Temperature and solar radiation are highly positively correlated, as the water will heat up when the sun shines and cool down when it does not, with $\rho_{23} = 0.75$. A weak negative correlation is observed between temperature and nutrient concentration, and between solar radiation and nutrient concentration, as phytoplankton absorb nutrients when they grow and release them to the water when they die, with $\rho_{12} = -0.1$ and $\rho_{13} = -0.4$. The equation describing the phytoplankton concentration, P, is given by

$$P(X_1, X_2, X_3) = a_1 X_1 + a_2 X_2 + a_3 X_3 - a_0 \tag{6.9}$$

where $a_1 = 0.05, a_2 = 0.08 \text{ mg/(m}^3°C)$, $a_3 = 0.01 \text{ mg/(mW)}$ and $a_0 = 1.5 \text{ mg/m}^3$. This is a great oversimplification, and the interested reader is directed to textbooks on this subject for further details (e.g., James 1993). The *failure criterion* is that the phytoplankton concentration exceeds a specific value that is considered harmful, say $P_0 = 12 \text{ mg/m}^3$. The *basic variables*, water temperature, nutrient concentration and solar radiation, are described by probability density functions, and therefore a *Level 2 method* of solution will be appropriate.

Solution. The reliability function may be written as

$$G = P_0 - (a_1 X_1 + a_2 X_2 + a_3 X_3 - a_0)$$
$$= P_0 + a_0 - a_1 X_1 - a_2 X_2 - a_3 X_3$$

Failure occurs when $G < 0$. Now, G is a linear function of the basic variables, which are normal with known correlation properties. The reliability index may be calculated using Equation (5.17)

Thus,

$$\beta = \frac{P_0 + a_0 - a_1 \mu_1 - a_2 \mu_2 - a_3 \mu_3}{\sqrt{\sum_{i=1}^{3} \sum_{j=1}^{3} a_i a_j \rho_{ij} \sigma_i \sigma_j}}$$

$$= \frac{12 + 1.5 - (0.05).100 - (0.08).16 - (0.01).130}{\sqrt{\begin{array}{l} 0.05^2 \times (75)^2 + 0.08^2 \times (6)^2 + 0.01^2 \times (70)^2 + (2) \times (0.05) \times (0.08) \times (-0.1) \times (75) \times (6) \\ + (2) \times (0.05) \times (0.01) \times (-0.4) \times (75) \times (70) + (2) \times (0.08) \times (0.01) \times (0.75) \times (6) \times (70) \end{array}}}$$

$$= \frac{5.92}{\sqrt{14.06 + 0.23 + 0.49 - 0.36 - 2.1 + 0.50}}$$

$$= \frac{5.92}{3.58}$$

$$= 1.654$$

The probability of failure is, therefore,

$$P_f = F_G(0) = \Phi(-1.654) = 1 - \Phi(1.654) \approx 1 - 0.951 = 0.049$$

The probability that the phytoplankton concentration exceeds the allowable value is 0.049, or about 5%.

Example 6.7. Scouring of a river bed

Statement of problem – consider a wide channel of uncertain cross-section, with $R \sim U(4, 8)$, where the slope and roughness are random variables with $S \sim U(0.0015, 0.0045)$ and $n \sim \text{Rayleigh}(0.03)$. Sediment transport occurs if the flow velocity exceeds the critical shear velocity V_c. However, this is known only imprecisely, and $V_c \sim \text{exponential}(1.5)$. Determine the probability that scour occurs.

Solution. It is assumed that the *basic variables*, V_c, S, n and R, are independent. The *failure criterion* is simply that the flow velocity exceeds the critical value. It would be possible, in principle, to determine the distribution function of the reliability function, and thence determine the probability of it being less than zero. Here, given that the reliability function is nonlinear and the variables are correlated and non-normal, an *approximate Level 2 method of solution* is used.

Now, $R \sim U(4, 8)$, and the roughness n and slope S are known. From Equation (6.2a),

$$V = \frac{S^{1/2} R^{2/3}}{n} \equiv A R^{2/3} \tag{6.10}$$

The reliability function may be written as

$$G = V_c - V = V_c - \frac{S^{1/2} R^{2/3}}{n} \tag{6.11}$$

which is a nonlinear function of non-normal variables. One could use an iterative Level 2 method, but here the use of the point estimation method (PEM) is illustrated.

G is a function of four random variables. Thus, values of G are required at $2^m \equiv 2^4 = 16$ points. Following Equations (5.50–5.53), these points are $z_k = G(\mu_1 + \varepsilon_{1k}\sigma_1, \mu_2 + \varepsilon_{2k}\sigma_2, \ldots, \mu_m + \varepsilon_{mk}\sigma_m)$, and are summarised in Table 6.1, with the obvious shorthand. The means of V_c, S, R and n are written as μ_1, μ_2, μ_3 and μ_4 etc.

Table 6.1 Points for PEM and corresponding values of G

$Z_1 = G(\mu_1 + \sigma_1, \mu_2 + \sigma_2, \mu_3 + \sigma_3, \mu_4 + \sigma_4)$	-2.256
$Z_2 = G(\mu_1 - \sigma_1, \mu_2 + \sigma_2, \mu_3 + \sigma_3, \mu_4 + \sigma_4)$	-3.590
$Z_3 = + - + +$	-2.256
$Z_4 = - - + +$	-3.589
$Z_5 = + + - +$	-1.322
$Z_6 = - + - +$	-2.656
$Z_7 = + - - +$	-1.322
$Z_8 = - - - +$	-2.655
$Z_9 = + + + -$	-10.415
$Z_{10} = - + + -$	-11.750
$Z_{11} = + - + -$	-10.412
$Z_{12} = - - + -$	-11.745
$Z_{13} = + + - -$	-7.358
$Z_{14} = - + - -$	-8.692
$Z_{15} = + - - -$	-7.356
$Z_{16} = - - - -$	-8.690

Table 6.2 Summary of basic variables

Variable	Distribution	Mean	Standard deviation
V_c	Exponential(0.6)	$1/1.5 = 0.6666$	0.6666
S	$U(0.0015, 0.0045)$	0.003	7.5×10^{-7}
R	$R \sim U(4, 8)$	6	1.3333
N	Rayleigh(0.03)	$0.03 \times (\pi/2)^{1/2} = 0.0376$	$0.03 \times \sqrt{[2-(\pi/2)]} = 0.020$

From Equation (5.52) the moments of G are estimated by

$$E(G^r) = \sum_{k=1}^{16} \theta_k z_k^r$$

where $\theta_k = 1/16$, as the basic variables are independent. To evaluate the z_k, the means and standard deviations of the basic variables are required. These are summarised in Table 6.2.

Using these, the values for z_k shown in the second column in Table 6.1 are obtained from direct calculation. Expressions for the mean and variance of G then follow as:

$$E(G) = \frac{1}{16}[-2.256 + (-3.590) + \ldots + (-8.690)] \qquad (6.12)$$

$$= -\frac{96.1}{16} = -6.00$$

$$\text{Var}(G) = \frac{1}{16}[(-2.256)^2 + \ldots + (-8.690)^2] - E(G)^2 \qquad (6.13)$$

$$= \frac{800}{16} - 36.0 = 14.0$$

The reliability index is thus $\mu/\sigma = -0.428$ and the probability of failure is $\Phi(-\beta) = \Phi(0.428) = 0.667$. The probability of failure is very high in this case, $\sim 67\%$, which means that scouring of the river bed is very likely.

6.3   Coastal flood defences

6.3.1   *Introduction*

Property and farmland close to the shoreline may be exposed to significant flood risk and have high value. Levels of flood protection vary from country to country. For example, in The Netherlands, where two-thirds of the country is below storm surge level, large rural areas may be protected by defences designed to withstand the 1 in 10,000 year event, with less densely populated areas protected to 1 in 4000 years. In the UK, where less of the country is below sea level, new residential developments are required to be defended to the 1 in 200 year level. Many design formulae for structures exposed to waves are underpinned by linear wave theory. This is also known as 'first-order theory' or 'Airy wave theory'. The essential assumption is that wave heights are small compared with wave length and water depth. Assume a sinusoidal free surface, $\eta(x,t)$, that is,

$$\eta = \frac{H}{2} \cos 2\pi \left(\frac{x}{L} - \frac{t}{T} \right) = \frac{H}{2} \cos (kx - \omega t) \tag{6.14}$$

and

$$L = \text{wave length} \qquad k = \frac{2\pi}{L} = \text{wave number}$$

$$T = \text{wave period} \qquad \omega = \frac{2\pi}{T} = \text{wave frequency}$$

Figure 6.6 shows the definition sketch for a sinusoidal wave and the definition of variables. Under the assumptions that the fluid is incompressible, inviscid and irrotational, the Navier-Stokes equations reduce to a Laplace field equation, which may be solved to find expressions for the horizontal and vertical components of velocity. The solution holds subject to the condition that:

$$\omega^2 = gk \tanh kh$$

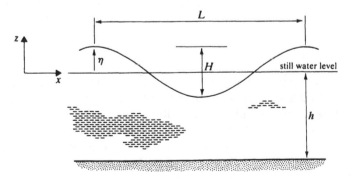

Figure 6.6 Definition sketch for a sinusoidal wave.

Table 6.3 Summary of linear wave equations

Wave parameter	General expression (transitional water depths)	Deep water $(d/L > 1/4)$	Shallow water $(d/L > 1/20)$
Surface profile (η)	$\eta = \dfrac{H}{2}\cos(kx - \omega t) = \dfrac{H}{2}\cos\theta$		
Celerity $(C = L/T)$	$C = \dfrac{g}{\omega}\tanh(kh)$	$= \dfrac{g}{\omega}$	$= \sqrt{gh}$
Length (L)	$L = \dfrac{gT}{\omega}\tanh(kh)$	$= \dfrac{gT}{\omega}$	$= T\sqrt{gh}$
Group velocity (C_g)	$C_g = nC$ $= \dfrac{C}{2}\left[1 + \dfrac{4\pi h/L}{\sinh 4\pi h/L}\right]$	$= \dfrac{C}{2}$	$= \sqrt{gh}$

This describes how the wave frequency is related to the wave number, and is known as the dispersion relation. The speed of wave propagation, or wave celerity (sometimes termed the phase speed), c, is given by:

$$c = \frac{L}{T} = \frac{\omega}{k} = \left(\frac{g}{k}\tanh kh\right)^{\frac{1}{2}} = \frac{g}{\omega}\tanh kh \qquad (6.15)$$

For deep water and shallow water, some simplification of these expressions is possible, and they are summarised in Table 6.3.

The potential energy is $E_p = \rho g H^2 L/16$ and kinetic energy is $E_k = \rho g H^2 L/16$ per unit crest length for one wave length. The total energy E per unit area is, therefore,

$$E = \frac{\rho g H^2}{8}$$

This can be large, that is, a force 8 gale after 24 hours can give a wave height of 5 m, corresponding to an energy in excess of $30\,\mathrm{kJ/m^2}$.

Wave transformations

The reader will need to be aware of three other important processes. These are refraction, shoaling and breaking. For detailed discussion the reader is referred to textbooks on the subject (e.g., Kamphuis 2001, Dean & Dalrymple 2002, Reeve et al. 2004). However, a brief explanation is given here. The process of *refraction* is the tendency of waves to turn so that their crests approach almost parallel to the shoreline. This happens because waves in deeper water travel faster than those in shallow water. If waves are approaching the shore obliquely, the section of their crest that is in deeper water travels faster, thereby making the crest line align more closely with the seabed contours. Waves can change their height through the process of *shoaling*. If a wave propagates into shallower water then, according to Airy theory, its speed reduces. If the rate at which the wave propagates its energy is conserved (i.e., there are negligible losses due to friction), then a reduction in wave speed must be counterbalanced by an increase in wave height, so that the energy-transmission rate remains unaltered. As waves propagate into shallower water, the process of shoaling leads to increasing wave heights. This cannot continue indefinitely, and eventually the wave *breaks*, dissipating energy through the production of turbulence, heat and sound. Waves break because their steepness (H/L) becomes very large as the depth decreases. The fluid velocity at the crest becomes large and the crest topples because it is unstable. The main limiting factors are:

1 Steepness $H/L < 1/7$
2 Ratio of height to depth (the breaking index) $\gamma = H/h = 0.78$

These criteria are for monochromatic (single frequency) waves rather than random waves, and should be used as guidance only in design, as in practice $0.4 \leq \gamma \leq 1.2$.

Figure 6.7 shows wave transformations at Bigbury Bay, Devon. As a further complication, waves can break in a number of different ways, which affects the forces they exert when impinging on a sea defence.

Figure 6.7 Wave transformations at Bigbury Bay, Devon, England (photograph courtesy of Dr S. M. White).

Figure 6.8 illustrates the main types of wave transformation that are common in coastal areas. In coastal situations, the primary limiting factor is water depth, and many design issues revolve around whether the design conditions should be defined by unbroken, breaking or broken waves. The type

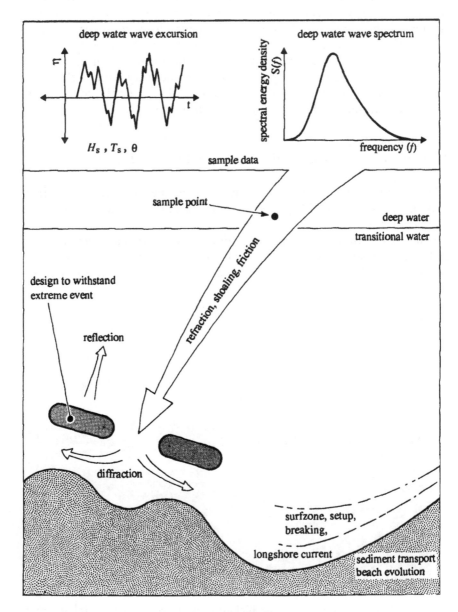

Figure 6.8 Wave transformations: main concepts.

of breaking may be determined approximately by the value of the Irribarren number (or surf similarity parameter):

$$\xi_b = \frac{\tan \beta}{\sqrt{\frac{H_b}{L_b}}} \tag{6.16}$$

where $\tan (\beta)$ is the beach slope, and H_b and L_b are the wave height and wave length at the breaking point. The breaker types are defined as follows (Figure 6.9):

Spilling $\quad \xi_b < 0.4$

Plunging $\quad 0.4 < \xi_b < 2.0$ (6.17)

Surging $\quad \xi_b > 2.0$

Any unidirectional sea state can be described mathematically as being composed of an infinite series of sine waves of varying amplitude and frequency. Thus, the surface excursion at any time $\eta(t)$ may be represented as

$$\eta(t) = \sum_{n=1}^{\infty} c_n \cos (\omega_n t + \phi_n)$$

where c_n are the amplitudes, ω_n are the frequencies and ϕ_n are phase angles. Figure 6.10 illustrates how such a summation of waves can build into a random-looking sea surface.

Note that, from the equation for wave energy ($E = \rho g H^2/8$), the wave energy is proportional to (amplitude)^2/2 (with units of m^2). Thus, the

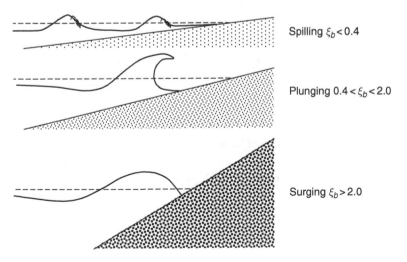

Spilling $\xi_b < 0.4$

Plunging $0.4 < \xi_b < 2.0$

Surging $\xi_b > 2.0$

Figure 6.9 Classification of breaking waves.

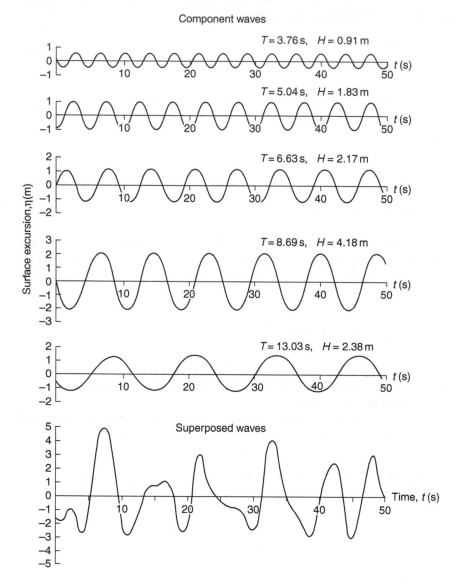

Figure 6.10 Superposition of harmonics to create a random wave train.

spectral energy density function $S(f)$ (with units of m²s), which is often expressed in terms of the frequency $f = (\omega/2\pi)$, may be found from

$$S(f)\Delta f = \sum_{f}^{f+\Delta f} \frac{1}{2}c_n^2 \tag{6.18}$$

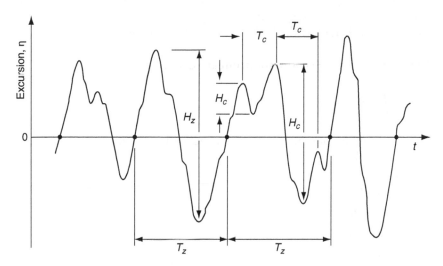

• Up-crossings of the mean level

Figure 6.11 Illustration for the definition of commonly used wave parameters.

Table 6.4 Approximate equivalence of some time and spectral domain
wave parameters

Time domain parameter	Equivalent frequency domain parameter
H_s	H_{m0} (approximate)
η_{rms}	$m_0^{0.5}$ (exact)
T_z	T_{m02} (approximate)
T_s	$0.95T_p$ (approximate)

Rather than dealing with idealised or empirical energy spectra, many
design formulae are cast in terms of quantities that may be derived from an
analysis of time series, and these have an approximate equivalent in terms
of moments of the energy spectrum. Figure 6.11 and Table 6.4 summarise
some of these quantities and the relationships between the time and spectral
domain parameters, where:

1   H_z (mean height between zero upward crossing);
2   T_z [mean period between zero upward (or downward) crossings];
3   H_c (mean height between wave crests);
4   T_c (mean period between wave crests);
5   H_{max} (maximum difference between adjacent crest and trough);
6   H_{rms} (root-mean-square wave height);

7   $H_{1/3}$ (mean height of the highest one-third of the waves), also known as the significant wave height or H_s;
8   $H_{1/10}$ (mean height of the highest one-tenth of the waves).

The moments of the energy spectrum are defined as:

$$m_n = \int_0^\infty S(f)f^n df$$

Spectral domain parameters that are defined in terms of the spectral moments include:

$$H_{m0} = 4(m_0)^{0.5}$$
$$T_{m01} = m_0/m_1$$
$$T_{m02} = (m_0/m_2)^{0.5}$$
$$T_p = 1/f_p$$

where f_p is the frequency at the maximum value of $S(f)$.

Frequency domain wave parameters do not have direct equivalent parameters in the time domain. However, as a guide, Table 6.4 summarises the approximate equivalences.

Types of structure

The cross-section of a typical breakwater is shown in Figure 6.12. This type of breakwater is often referred to as a 'rubble mound' breakwater, as the core, filter and armour layers are composed of randomly placed material. Its primary aim is to limit wave action and overtopping. The main elements are the primary armour, which will be designed to withstand the forces of

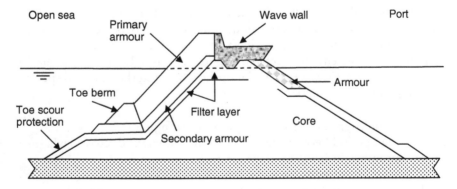

Figure 6.12 Typical elements of a port rubble-mound breakwater.

incoming storm waves, and the toe protection, which prevents erosion of the seabed and undermining of the structure.

Other forms of breakwater are the caisson type, shown in Figure 6.13, and the solid stone/concrete type, an example of which is shown in Figure 6.14.

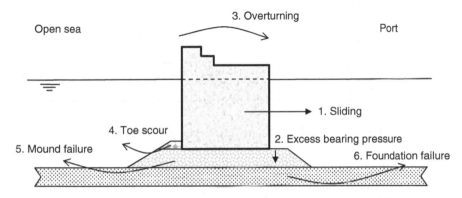

Figure 6.13 Illustrative vertical caisson-type breakwater with primary failure modes.

Figure 6.14 Wave overtopping of a stone/concrete near-vertical seawall at Aberystwyth, Wales. Note also the return flow of the overtopped water through the pipe outflow in the wall towards the left-hand end of the picture (photograph courtesy of Dr Adrián Pedrozo-Acuña).

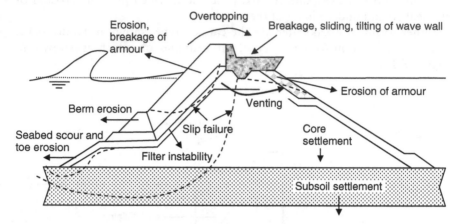

Figure 6.15 Overview of rubble-mound breakwater failure modes.

As noted in Chapter 1, the failure of the breakwater at Sines, amongst others, prompted much research into the failure modes of breakwaters. Figure 6.15 shows some of the main failure modes of rubble mound breakwaters. Here, two of these are considered in more detail: wave overtopping and the ability of the primary armour to withstand wave action.

Of the many formulae developed over the years for defining the size of the armour units (i.e., to ensure the hydraulic stability of the armour layer), the most enduring have been those due to Hudson (1959) and Van der Meer (1988). Hudson developed his formula by laboratory testing with regular (monochromatic) waves. The formula is:

$$D_{n50}^3 = \frac{H^3}{K_D(\rho_s/\rho_w - 1)^3 \cot(\alpha)} \tag{6.19}$$

where:

H = characteristic wave height (often taken as H_s when random waves are considered)
D_{n50} = equivalent cube length of median rock
ρ_s = density of rock
ρ_w = density of water
α = slope angle of the armour facing the waves

The quantity $(\rho_s/\rho_w - 1)$ is often written as Δ. Although really superseded today by the Van der Meer equations, the Hudson formula is still a useful

Table 6.5 K_D values for armour on the trunk of a breakwater

Unit	Layer thickness in number of units	Breaking waves	Nonbreaking wave
Quarrystone			
Rounded	2	1.2	2.4
Angular	2	2.0	4.0
Angular	>3	2.2	4.5
Tetrapod	2	7.0	8.0
Dolos	2	15.8*	31.8*

* This refers to <5% rocking.

tool for non-extreme conditions, slopes that are within the range which the original model tests validated (generally slopes of 1:2 or 1:3 and, in some circumstances, 1:1.5), and also where generally conservative values are adopted for the K_D factors. The strong influence of the armour unit density is apparent, being within the expression that is raised to the cube power. Table 6.5 provides recommended K_D values for different types of unit from CERC (1984).

Van der Meer's equations, which were developed by means of a series of model tests carried out at the Delft Hydraulics Laboratory, have become the current 'standard' – but are more complicated to apply. Van der Meer published his results in a monograph entitled 'Stability of Cubes, Tetrapods and Accropodes, Design of Breakwaters', but has since published many more papers on stability, wave run-up and transmission. A very comprehensive list of references is given at the end of Section VI of the CEM (see references). Van der Meer's formula for two-layered armour on non-overtopped slopes is, for plunging waves (defined as $\xi_m < \xi_{mc}$):

$$\frac{H_s}{(\rho_s/\rho_w - 1)^3 D_{n50}} = 6.2 S_d^{0.2} P^{0.18} N_z^{-0.1} \xi_m^{-0.5} \tag{6.20}$$

and for surging waves, with $\xi_m > \xi_{mc}$

$$\frac{H_s}{(\rho_s/\rho_w - 1)^3 D_{n50}} = 1.0 S_d^{0.2} P^{-0.13} N_z^{-0.1} [\cot(\alpha)]^{0.5} \xi_m^P \tag{6.21}$$

where $\xi_m = s_m^{-0.5} \tan\alpha$ and $\xi_{mc} = \{6.2 P^{0.31} (\tan\alpha)^{0.5}\}^{1/(P+0.5)}$

and:

H_s = significant wave height in front of the breakwater.
D_{n50} = equivalent cube length of median rock ($M_{50} = \rho_s D_{n50}^3$)
ρ_s = density of rock or concrete unit
ρ_w = density of water
S_d = relative eroded area or damage level
P = notional permeability
N_z = number of waves
α = slope angle of the seaward facing slope
s_m = wave steepness = H_s / L_{om}
L_{om} = deepwater wave length corresponding to mean wave period.

Typical ranges for parameters P, S_d and rock density are: $0.1 \leq P \leq 0.6, 0.005 \leq S_d \leq 0.06$ and $2000\,\text{kg/m}^3 \leq \rho_s \leq 3100\,\text{kg/m}^3$.

Overtopping formula

The estimation of wave overtopping volume is important, not just for harbour breakwaters, but also for coastal flood defences.

Figure 6.16 shows a typical arrangement of a coastal flood defence, together with definitions of some of the most important variables. The formula used in the UK for many years, based on laboratory experiments by Owen (1980) is:

$$Q_m = Q^* T g H_s \tag{6.22}$$

where $R^* = R_c / [T_m (g H_s)^{0.5}]$, $Q^* = A \exp(-B R^* / r)$, and A and B are empirically derived coefficients given in Table 6.6 (for straight slopes only).

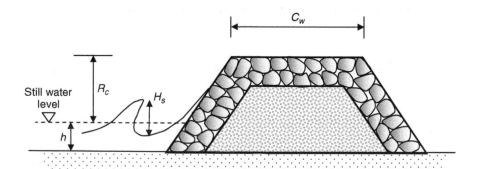

Figure 6.16 Definition of terms for wave overtopping. The still water depth is h, the significant wave height H_s, the crest width C_w and the freeboard R_c.

Table 6.6 Owen's coefficients for simple slopes

Seawall slope	A	B
1:1	7.94×10^{-3}	20.1
1:1.5	8.84×10^{-3}	19.9
1:2	9.39×10^{-3}	21.6
1:2.5	1.03×10^{-3}	24.5
1:3	1.09×10^{-2}	28.7
1:3.5	1.12×10^{-2}	34.1
1:4	1.16×10^{2}	41.0
1:4.5	1.20×10^{-2}	47.7
1:5	1.31×10^{-2}	55.6

This formula appears in the recently published EurOtop manual, which presents an integrated set of results for overtopping drawn from across Europe (EurOtop 2008).

6.3.2 Examples

Example 6.8. Tsunami warning

Statement of problem – For a tsunami warning system there is a need to estimate wave travel times from an earthquake region to a distant shoreline. The seabed levels are known reasonably well, but with some uncertainty. The mean undisturbed water depth that a tsunami wave experiences while propagating to a distant coast is 40 m. Variations about this depth due to undulations in the seabed may be described by a uniform distribution $U(-10, 10)$. The distance travelled by the tsunami from its point of generation to the coast is 4000 km. Determine the distribution of the tsunami propagation speed, and thus the uncertainty in arrival time. No *failure criterion* is specified, but the result will provide a measure of uncertainty in estimates of the arrival time of the tsunami, and hence the amount of time available for warning and preparation. The *basic variable* is water depth, which is described by a probability density function. The *method of solution* will be to calculate the probability density function of a transformed variable.

Solution. Tsunami waves have a very long wave length (many kilometres), so in the case above they may be considered to be shallow-water waves as their wave length is much greater than the water depth. To a good degree of approximation, it may be described as a shallow-water wave with wave speed, $c = \sqrt{(gh)}$, where h is the water depth. The water depth may be described by a $U(30, 50)$ distribution. The problem is to determine

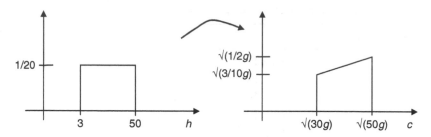

Figure 6.17 Illustration of the transformation of the water depth density function to the wave speed density function.

the probability density of c, given the probability density of h, and the relationship between c and h. Referring to Example 3.5, we see that

$$f_c(c) = \frac{f_h\left(\dfrac{c^2}{g}\right)}{\sqrt{g/4h}} = \begin{cases} 2\dfrac{c}{g}f_h\left(\dfrac{c^2}{g}\right) & \sqrt{30g} \le c \le \sqrt{50g} \\ 0 & \text{otherwise} \end{cases}$$

but $f_h(c^2/g)$ is just a uniform distribution so the distribution of c is actually linearly increasing over the given range. Figure 6.17 illustrates the transformation of distributions. The result of the change in distribution means that, although the tsunami speed obtained by using the midpoint of the range of possible speeds is $\{\sqrt{(50g)} + \sqrt{(30g)}\}/2 = 19.66$ m/s, the majority of the area under the probability density function (and thus the probability of occurrence) lies slightly to the right of this value. The median value of the tsunami speed (the value of c for which the integral of its density function equals 0.5) is given by $\sqrt{(40g)} = 19.81$ m/s. The propagation time using the two speeds, respectively, is $4,000,000/19.66 = 56.52$ hours and $4,000,000/19.81 = 56.09$ hours. As the distribution is skewed, using the mean value rather than the median value in this case results in a smaller value of the speed and thus travel time. The difference is not great, but in the case of a tsunami, an extra half an hour could be extremely valuable in terms of saving people and property. When dealing with skewed distributions, using the median rather than the mean provides a 'fairer' representation of the 'typical' value, as there is a 50% chance that a value obtained at random will be smaller than the median.

Example 6.9a–d. This group of examples covers a rock armour stability problem, a variation on this, and solutions of these obtained with iterative and PEM methods in both cases. The initial problem follows the development given by Burcharth (1992).

a   *Statement of problem* – It is required to assess the stability of a breakwater rock armour layer using Hudson's formula, where $K_D = 4, \cot(\alpha) = 2$ and $\Delta = 1.6$. These variables are taken to be exact, but uncertainties in them are accounted for by a single multiplicative factor A, which has an $N(1, 0.18)$ distribution. It is also given that $H_s \sim N(4.4, 0.7)$ and $D_n \sim N(1.5, 0.1)$, and that the variables are independent. The *failure criterion* is that the rock size is no longer able to maintain stability against the wave conditions. Rewriting Hudson's formula as a reliability function yields:

$$G = AD_{n50}\Delta[K_D \cot(\alpha)]^{1/3} - H_s$$

The *basic variables* are A, D_{n50} and H_s. As these variables are described by probability distributions, a *Level 2 approach* is appropriate. The reliability function is a nonlinear function of the basic variables, so an iterative method will be used.

Solution. The failure surface corresponding to the reliability function is, for $K_D = 4$,

$$1.59AD_{n50}\Delta \cot(\alpha)^{1/3} - H_s = 0$$

Using the transformation (Equation 5.16), and writing X_1 for A, X_2 for D_{n50} and X_3 for H_s, the failure surface in the normalised coordinate system is given by:

$$(1 + 0.18z_1) \times 1.6 \cdot (1.5 + 0.1z_2) \times 2^{1/3} \times 1.59 - (4.4 + 0.7z_3) = 0$$

or

$$0.864z_1 + 0.32z_2 + 0.058z_1z_2 - 0.7z_3 + 0.4 = 0$$

One of the alternative iterative solution methods is now illustrated. Substituting $z_i = \beta\alpha_i$ in the above, following the Hasofer–Lind definitions in Chapter 5, gives

$$\beta = \frac{-0.4}{0.864\alpha_1 + 0.32\alpha_2 + 0.058\alpha_1\alpha_2\beta - 0.7\alpha_3}$$

By use of Equations (5.21)–(5.29):

$$\alpha_1 = -\frac{(0.864 + 0.058\beta\alpha_2)}{K}$$

$$\alpha_2 = -\frac{(0.32 + 0.058\beta\alpha_1)}{K}$$

$$\alpha_3 = \frac{0.7}{K}$$

and

$$K = \sqrt{(0.864 + 0.058\beta\alpha_2)^2 + (0.32 + 0.058\beta\alpha_1)^2 + (0.7)^2}$$

The iteration is now initiated by choosing starting values for $\beta, \alpha_1, \alpha_2$, and α_3, and then calculating new values until convergence is achieved. This process is shown in Table 6.7 below. Convergence tends to be faster if α values related to loading variables are positive and those relating to resistance variables are negative.

The probability of failure is then

$$P_f = \Phi(-\beta) = \Phi(-0.341) = 0.367$$

The design point coordinates in the normalised z-coordinate system are $(z_1^d, z_2^d, z_3^d) = (\beta\alpha_1, \beta\alpha_2, \beta\alpha_3) = (-0.255, -0.091, 0.208)$. Inverse transforming these back to the original 'X'-coordinates gives $(A^d, D_{n50}^d, H_s^d) = (0.954, 1.491, 4.546)$. The α values give the relative importance of the random variables to the failure probability. Table 6.8 shows that the uncertainty related to the stone size is of relatively little importance in comparison to the uncertainties in wave height and the uncertainty factor A.

In itself, this result suggests that most effort should be put into the careful estimation of the wave heights, and assessing the suitability of the underlying stability formula.

b *Statement of problem* – With exactly the same conditions as in Example 6.9a, estimate the probability of failure using PEM. The *failure criterion* is exactly the same, as are the *basic variables*. The *method of*

Table 6.7 Iteration of variables for Example 6.9a

	Iteration number			
	Initial value	1	2	3
β	3.0	0.438	0.342	0.341
K		1.144	1.149	1.149
α_1	-0.50	-0.744	-0.747	-0.747
α_2	-0.50	-0.263	-0.266	-0.266
α_3	0.50	0.612	0.609	0.609

Table 6.8 Relative importance of basic variables
to the probability of failure

i	Variable (X_i)	α_i	$(\alpha_i)^2$ (%)
1	A	-0.747	55.8
2	D_n	-0.266	7.1
3	H_s	0.609	37.1

solution is specified as PEM. There are three basic variables, which are
independent, and hence the three-variable version of the PEM algorithm
suitable for independent variables is used.

Solution. G is a function of three random variables, and hence evaluation at $2^m \equiv 2^3 = 8$ points will be required. Following Equations
(5.50)–(5.53), these points are $z_k = G(\mu_1 + \varepsilon_{1k}\sigma_1, \mu_2 + \varepsilon_{2k}\sigma_2, \mu_3 + \varepsilon_{3k}\sigma_3)$,
and are summarised in Table 6.9, with the obvious shorthand. Also, the
means of A, D_{n50} and H_s have been written as μ_1, μ_2 and μ_3 etc.
From Equation (5.52) the moments of G are estimated by

$$E(G^r) = \sum_{k=1}^{16} \theta_k z_k^r$$

where $\theta_k = 1/8$, as the basic variables are independent. Estimates of the
mean and variance of G are then obtained as follows:

$$E(G) = \frac{1}{8}[0.951 + (-0.895) + \ldots + (-0.021)] = \frac{3.26}{8} = 0.408$$

$$\mathrm{Var}(G) = \frac{1}{8}[(0.951)^2 + \ldots + (-0.021)^2] - E(G)^2$$

$$= \frac{12.089}{8} - 0.166 = 1.345$$

Table 6.9 Points for PEM in Example 6.9b

$Z_1 = G(\mu_1 + \sigma_1, \mu_2 + \sigma_2, \mu_3 + \sigma_3)$	0.951
$Z_2 = G(\mu_1 - \sigma_1, \mu_2 + \sigma_2, \mu_3 + \sigma_3)$	-0.895
$Z_3 = + - +$	0.195
$Z_4 = - - +$	-1.421
$Z_5 = + + -$	2.351
$Z_6 = - + -$	0.505
$Z_7 = + - -$	1.595
$Z_8 = - - -$	-0.021

The reliability index is thus $\mu/\sigma = 0.303$ and the probability of failure is $\Phi(-\beta) = 1 - \Phi(0.303) = 0.382$. This is very close to the value computed using the iterative method in part (a) of this example. This method is also much quicker to evaluate, and does not require iteration. However, it gives neither the design point nor any indication of which variables are contributing most to the probability of failure.

c   *Statement of problem* – We have exactly the same conditions as in Example 6.9a, except that the distribution function for the wave heights is taken to be a Gumbel distribution, with the same mean and standard deviation as before. This is more realistic than the assumption of normality, as for storm conditions wave heights may be expected to follow an extreme distribution. The *failure criterion* is exactly the same, as are the *basic variables*. The basic variables are all described by probability density functions, so a *Level 2 approach* is suitable. The difference from Example 6.9a is that one of the basic variables is not normal. Therefore, the *method of solution* will require a modified version of the iterative technique.

Solution. The Gumbel distribution is given in Equation (4.5). The values of the mean and standard deviation, 4.4 m and 0.7 m, respectively, determine the parameters of the Gumbel distribution as $\mu = 4.08$ m and $\sigma = 0.546$ m. These are required later in the iteration to transform from the Gumbel to the normal distribution. The working now proceeds as before, but with a modified mean and variance for the wave height variable. The failure surface in the normalised coordinate system is given by:

$$(1 + 0.18z_1) \times 1.6 \times (1.5 + 0.1z_2) \times 2^{1/3} \times 1.59 - (\mu'_{x_3} + \sigma'_{x_3} z_3) = 0$$

or

$$0.864z_1 + 0.32z_2 + 0.058z_1z_2 - \sigma'_{x_3} z_3 + (4.8 - \mu'_{x_3}) = 0$$

Substituting $z_i = \beta \alpha_i$ in the above, gives

$$\beta = \frac{-(4.8 - \mu'_{x_3})}{0.864\alpha_1 + 0.32\alpha_2 + 0.058\alpha_1\alpha_2\beta - \sigma'_{x_3}\alpha_3}$$

By use of Equations (5.21)–(5.29):

$$\alpha_1 = -\frac{(0.864 + 0.058\beta\alpha_2)}{K}$$

$$\alpha_2 = -\frac{(0.32 + 0.058\beta\alpha_1)}{K}$$

$$\alpha_3 = \frac{\sigma'_{x_3}}{K}$$

and

$$K - \sqrt{(0.864 + 0.058\beta\alpha_2)^2 + (0.32 + 0.058\beta\alpha_1)^2 + (\sigma'_{x_3})^2}$$

From Equation (5.32),

$$x_3^d = F_G^{-1}[\Phi(\beta\alpha_3)]$$

and from Equations (5.33) and (5.34)

$$\sigma'_{x_3} = \frac{\phi[\Phi^{-1}(F_G(x_3^d))]}{f_G(x_3^d)}$$

and

$$\mu'_{x_3} = x_3^d - \Phi^{-1}[F_G(x_3^d)] \times \sigma'_{x_3}$$

The iteration now proceeds as follows. First, choose starting values for $\alpha_1, \alpha_2, \alpha_3$ and β. Next, find x_3^d, and then μ'_{x_3} and σ'_{x_3}. Then find the new value of β, and update α_1, α_2 and α_3, and so on, until convergence is achieved. The results of this procedure are shown in Table 6.10.

The probability of failure is then

$$P_f = \Phi(-\beta) = \Phi(-0.457) = 0.324$$

The design point coordinates in the normalised z-coordinate system are $(z_1^d, z_2^d, z_3^d) = (\beta\alpha_1, \beta\alpha_2, \beta\alpha_3) = (-0.342, -0.12, 0.277)$. Inverse transforming these back to the original 'X'-coordinates gives

Table 6.10 Iterations for Example 6.9c

	Iteration number							
	Initial value	1	2	3	4	5	6	7
β	3.0	1.717	0.553	0.569	0.463	0.461	0.457	0.457
K		1.295	1.363	1.165	1.155	1.144	1.143	1.143
α_1	−0.5	−0.629	−0.629	−0.735	−0.742	−0.749	−0.749	−0.750
α_2	−0.5	−0.199	−0.220	−0.254	−0.260	0.262	−0.262	−0.263
α_3	0.5	0.772	0.754	0.627	0.619	0.609	0.608	0.607
x_3^d		5.359	4.568	4.525	4.475	4.471	4.469	4.469
σ'_{x_3}	1.0	1.027	0.731	0.715	0.697	0.695	0.694	0.694
μ'_{x_3}	3.0	4.033	4.139	4.264	4.270	4.275	4.276	4.276

$(A^d, D_{n50}{}^d, H_s^d) = (0.934, 1.474, 4.468)$. In this case, the change in the probability density function for wave heights has not altered the probability of failure very much, reducing by only a few per cent.

d   *Statement of problem* – With exactly the same conditions as in Example 6.9c, estimate the probability of failure using PEM. The *failure criterion* is exactly the same, as are the *basic variables*. The *method of solution is specified as PEM*. We have three basic variables, which are independent, and hence the three-variable version of the PEM algorithm suitable for independent variables is used.

Solution. The PEM requires evaluation of the reliability function G at specific points defined by the mean and standard deviation of the basic variables. As the means and standard deviations of all the variables are the same as in Example 6.9b, the answer is the same, that is, the probability of failure is $\Phi(-\beta) = 1 - \Phi(0.303) = 0.382$.

Example 6.10. *Statement of problem* – Derive an expression for the probability of failure of a rock armour revetment using Van der Meer's formula. Use this to calculate, using the MVA method, the probability of failure for the specific conditions given below. Assume that $a, b, \Delta, S_d, H_s, T_m$ and D_{n50} are independent random variables with $a = N(6.2, 0.62)$, $b = N(0.18, 0.02)$, $\Delta = N(1.59, 0.13)$, $H_s = N(3, 0.3)$, $D_{n50} = N(1.30, 0.03)$, $S_d = 8, \cot(\theta) = 2.0$, and the number of waves equal to 2000. The *failure criterion* is that the wave conditions cause unacceptable damage to the revetment. The *basic variables* are a and b. The formulae parameters are Δ, the relative density; S_d, the relative eroded area; H_s, the significant wave height; T_m, the mean wave period; and D_{n50}, the nominal rock diameter. These are defined by probability distributions; therefore a *Level 2 method of solution* will be appropriate.

Solution. The response function is taken to be Van der Meer's (1988a) formula for armour stability under deep-water plunging waves. For a given damage level S_d, the formula provides an estimate of the required nominal median stone size D_{n50}. The failure function may be written as

$$G = R - S = aP^b S_d^{0.2} \Delta D_{n50} \sqrt{\cot(\alpha)} \left(\frac{g}{2\pi}\right)^{-\frac{1}{4}} - H^{0.75} \frac{0.5}{T_m^{0.5}} N^{0.1}$$

where H_s is the significant wave height, T_m is the mean wave period, Δ is the relative mass density $(\rho_s - \rho_w)/\rho_w, a = 6.2$ and $b = 0.18$, with some uncertainty defined by their probability density functions. The first step is to calculate the partial derivatives. Performing this analytically gives:

$$\frac{\partial G}{\partial a} = R/a$$

$$\frac{\partial G}{\partial H_s} = -0.75S/H_s$$

$$\frac{\partial G}{\partial T_m} = -0.5S/T_m$$

$$\frac{\partial G}{\partial D_{n50}} = R/D_{n50}$$

$$\frac{\partial G}{\partial b} = bR/P$$

$$\frac{\partial G}{\partial \Delta} = R/\Delta$$

Table 6.11 summarises the results of the MVA calculations, where values are quoted to two decimal places and hence there may be small discrepancies. For example, for the variable $a(0.62 \times 2.63)^2 = 2.65$, rather than the value 2.64 in the table.

Now,

$$\mu_G = \mu_R - \mu_s = 4.29$$

and

$$\sigma_G^2 = \sum_{i=1}^{6} \alpha_i^2 = 9.64$$

so $\sigma_G = 3.11$

Therefore, the reliability index $\beta = \dfrac{\mu_G}{\sigma_G} = 1.383$

and the probability of failure is $\Phi(-\beta) = 0.034$

In this relatively simple case the probability of failure has been calculated taking into account uncertainty in the parameter values of an empirical

Table 6.11 MVA results for Example 6.10

Variable	Mean	Standard deviation	Partial derivative	α_i^2
a	6.2	0.62	2.63	2.64
b	0.18	0.02	29.2	0.34
Δ	1.59	0.13	10.2	1.76
D_{n50}	1.30	5.00	12.5	0.14
T_m	6.0	2.0	-1.0	3.96
H_s	3.0	0.30	-3.0	0.80

equation, construction materials and the random nature of waves. In passing, it is interesting to note that the result is sensitive to the rock size and density, not just the wave conditions.

Example 6.11. *Statement of problem* – Given 20 years of 3-hourly measurements of water level, wave height, wave period and wave direction, determine using PEM the probability distribution of wave overtopping of a simple seawall with a front-face slope of 1 in 2, a roughness of 0.85 corresponding to stone blockwork, a crest of 10 m, a foreshore of a plane beach with a slope of 1 in 50 rising to a level of 1 m at the base of the structure. The *failure criterion* is not specified, only that the probability density function of overtopping is required. The *basic variables* are wave height, wave period, wave direction, water level, crest level, wall slope and wall roughness. Various *methods of solution* are possible, but here the results of a simulation–structure function approach are described. It is also required that PEM be used, but this does not provide a probability distribution function by itself, so an additional step to link the moments obtained from PEM to a probability density function is required.

Solution. Using Owen's equation for overtopping, Equation (6.22) may be recast in the following form for the mean overtopping rate:

$$Q_m = AT_m gH_s \exp\left[-BR_c/(T_m(gH_s)^{0.5})/r\right] \tag{6.23}$$

R_c is the freeboard, or the difference between the crest level and the water level. In principle, all variables in this equation, including g, could have uncertainty associated with them. The problem is thus to find the probability distribution of a nonlinear function of six random variables, which are probably non-normal and possibly correlated. This is a very difficult problem, and to make progress either simplifying assumptions or numerical simulation is necessary. One possibility is to use the time series data to calculate a time series of overtopping volumes, corresponding to each set of water level and wave conditions in the given series. This sequence of overtopping values can then be analysed using a univariate extremes analysis and a distribution function with best-fit parameters obtained. This approach has the advantage of side-stepping any issues of correlation, but requires specific values of the parameters A and B to be used. The analysis could be repeated many times over with differing values of A and B to test the sensitivity of results to uncertainties in these parameters.

Alternatively, marginal extremes analysis of the water levels and wave heights, combined with suitable assumptions about the wave period and direction could be used to perform a Level 2 type analysis. PEM could also be used in this case, although extensive preparatory analysis to determine the means and variances of the basic variables would be required, together

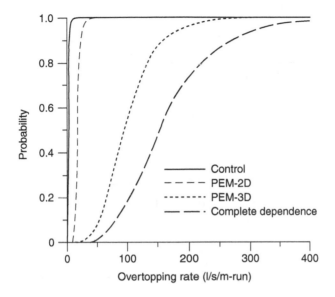

Figure 6.18 Cumulative probability curves of overtopping computed from a 30-year
synthetic time series data using: (i) Weibull fit to data; (ii) PEM applied
to extreme wave heights and water levels; (iii) PEM applied to extreme
wave heights, periods and water levels; (iv) the assumption of complete
dependence between wave heights and water levels.

with an estimation of intercorrelations. Figure 6.18 illustrates a comparison of various Level 2 methods and a Level 3 simulation for the case of wave overtopping of a simple seawall. The plots show the distribution function of overtopping discharge as determined using different assumptions. The assumption of complete dependence between waves and water levels provides an upper bound (i.e., the largest waves always occur with the largest water levels). The Level 3 result, obtained by generating a time series of overtopping rates from the time series of waves and water levels and then performing a univariate extremes analysis on the series, provides the least conservative result (labelled 'control' in Figure 6.18). Distributions derived using PEM (two-dimensional using wave height and water level, and three-dimensional using wave height, wave period and water level) lie between them. Further details may be found in Reeve (1998, 2003) and references therein. In passing, it is worth noting that the uncertainty represented by the different approaches does not provide a designer with much guidance. For example, taking an overtopping threshold of 100 l/s/m-run, the control curve and the PEM two-dimensional curve both show the probability of the overtopping being below this threshold as being close to or at unity. In contrast, the three-parameter PEM and complete dependence show probabilities of ~0.6 and 0.2, respectively.

Table 6.12 Details of distributions of basic variables in Example 6.12

Basic variable	Distribution	Mean	Standard deviation	(Percentage of mean)
Significant wave height(m)	Normal	3.0	10	
Slope angle (°)	None	0.5	–	
Rock density kg/m³	Normal	2,650	5	
Nominal rock diameter(m)	Normal	1.3	5	
Permeability parameter	None	0.1	–	
Wave steepness	Normal	0.05	10	
Van der Meer parameter *a*	Normal	6.2	10	
Van der Meer parameter *b*	Normal	0.18	10	

Example 6.12. Damage to a rock armour structure (adapted from Meadowcroft et al. 1995)

Statement of problem – It is required to estimate the probability of excessive damage to a rock armour structure using Van der Meer's equation. Uncertainty in the performance of the structure arises from sources such as variability in rock armour size, errors in estimating design wave height, and the approximate empirical nature of the design equation. For this example we take the distribution functions of the basic variables to be known and to be normal; these are given in Table 6.12. In practice, the choice of distributions and their parameters should be estimated against observations.

The *failure criterion* is that the degree of damage S_d does not exceed a value of 2, that is, minor damage only. The *basic variables* are listed in the first column of Table 6.12. Various methods of solution are possible. In this case, a Monte Carlo simulation is performed.

Solution. Given the probability density functions in Table 6.12, a Monte Carlo simulation was set up to generate 6000 values of each of the basic variables. In the absence of other information, the variables are treated as being independent. For each of the 6000 sets of values of the basic variables, the damage levels are calculated using Van der Meer's formula (Equation 6.20). The 6000 results of these calculations are then used to create a frequency plot and thus an empirical probability density function. This is then used to generate an exceedance plot (which plots 1 minus the cumulative distribution function against S_d), which is shown in Figure 6.19.

The ensemble of results have a mean of $S_d = 3$ and a standard deviation of 3.7. The probability distribution shows the predicted damage for a structure

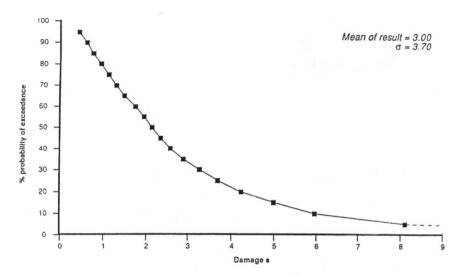

Figure 6.19 Probability of exceedance of the predicted damage for a structure designed for minor damage ($S_d = 2$).

designed for minor damage ($S_d = 2$), as a probability of exceedance. For example, the probability of the damage exceeding 6 is about 10%.

6.4   Flow in pipes and drains

6.4.1   *Introduction*

In this section, problems concerning the flow of fluid in pipes, tanks and drains are covered. A brief summary of some key equations is given, as well as introductory material on turbulent flow in pipes. The principle behind most elementary fluid flow problems is that of energy conservation and mass conservation. The first is often expressed as Bernoulli's equation for a unit weight of fluid:

$$z_1 + \frac{V_1^2}{2g} + \frac{p_1}{\rho g} = z_2 + \frac{V_2^2}{2g} + \frac{p_2}{\rho g} + \text{energy head losses} \tag{6.24}$$

In ideal flow, the energy head losses are zero, but in practice these are often specified by empirical formulae. In the above equation the indices refer to two particular points in the fluid, and z_i, V_i and p_i are the height above a fixed datum, velocity of the fluid, and pressure, respectively, at point i.

Water is conveyed from its source to treatment plants and then to consumers, usually in pressurised pipes and thence, via unpressurised drains, to

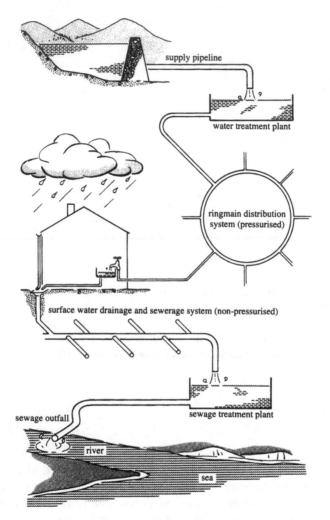

Figure 6.20 The synthetic hydrological cycle (from Chadwick and Morfett 1999).

sewage treatment plants to return to the rivers and seas through outfalls, see Figure 6.20.

At the treatment plants, and also to a lesser degree in drains and storm water containment tanks, a balance between the rate at which water flows in and the rate at which it flows out must be maintained. If the inflow exceeds the outflow by too much, the spare capacity in the tank/plant will be used up and flooding will occur. Conversely, if the outflow is greater than the inflow, the tank/plant will run dry.

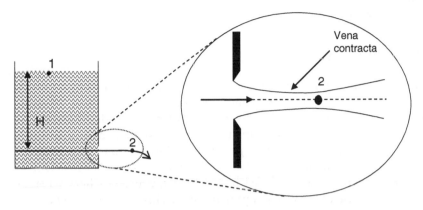

Figure 6.21 Discharge through an orifice. Left-hand panel – definition of terms for flow through an orifice in the side of a tank discharging freely to the atmosphere. Right-hand panel – detail of the flow as it leaves the tank, illustrating the vena contracta caused by the contraction of the streamlines as they pass through the plane of the orifice.

At its simplest, a storage tank may be considered to be open to the atmosphere, with drainage achieved through an orifice near the bottom of the tank. Bernoulli's equation can be applied to this situation at points 1 and 2 in Figure 6.21.

Neglecting energy losses, we assume the tank is large and the orifice is small so that, to a good degree of approximation, $V_1 = 0$. If we take atmospheric pressure as our reference pressure, then without loss of generality we can set $p_1 = 0$, and so $p_2 = 0$. Similarly, taking the level of point 2 to be the height datum, we have $z_2 = 0$ and $z_1 = H$. Consequently,

$$V_2 = \sqrt{2gH} \tag{6.25}$$

By conservation of mass, the discharge from the orifice Q_T should be

$$Q_T = A_T\sqrt{2gH} \tag{6.26}$$

where A_T is the area of the orifice. This equation does not completely reflect reality because the area of the jet at the vena contracta (see Figure 6.21), A, is less than the area of the orifice A_T. The area of the jet is a function of the roughness and geometry of the orifice, and is usually described by a coefficient of contraction, $C_C = A/A_T$. This is inserted as a multiplicative constant on the right-hand side of Equation (6.26). Even with this correction, an accurate prediction of the discharge is not obtained because we have assumed there is no energy loss. In practice, the nature of the flow through the orifice generates turbulence, which dissipates energy. A coefficient of velocity, $C_V = V_J/V_2$, is introduced to account for this, so that the actual discharge Q is given by

$$Q = C_V C_C A \sqrt{2gH} \tag{6.27}$$

Research into flow in pipes has a long history and is marked by a tension between those wishing to take a fluid mechanics view and those adopting a more empirical approach. The formulae that are in widest current use essentially provide a means of estimating the flow velocity as a function of the hydraulic radius, the friction slope (the head lost due to friction in a pipe of a given length), the nature of the flow and various coefficients. One of the earliest formulae was proposed by the French engineer Chézy when he was designing a canal for water supply to Paris. This may be written:

$$V = C\sqrt{RS} \tag{6.28}$$

where C is known as the Chézy coefficient, R is the hydraulic radius and S is the friction slope. A refinement of this formula related the Chézy coefficient to the hydraulic radius. This became known as the Manning equation, after the Irish engineer Robert Manning. In this equation $C = R^{1/6}/n$, where n is a coefficient known as Manning's n. Equation (6.28) thus reads

$$V = \frac{S^{1/2} R^{2/3}}{n} \tag{6.29}$$

Manning's equation is generally considered to be applicable for fully turbulent flow (i.e., with $Re > 4000$). The Darcy–Weisbach equation is a phenomenological formula developed by the French scientist Henri Darcy, and further refined into the form used today by Julius Weisbach of Saxony in 1845. It relates the head loss due to friction h_f along a given length of pipe L to the average velocity of the fluid flow:

$$h_f = \frac{\lambda L V^2}{2gD}$$

where D is the pipe diameter. This can be recast as an equation for velocity:

$$V = \sqrt{\frac{8gRS}{\lambda}} \tag{6.30}$$

where the friction slope $S = h_f/L$ and for full pipe flow the hydraulic radius $R = D/4$. The friction factor λ (often written as f in US texts) is not a constant but depends upon the Reynolds number and the hydraulic radius.

In the 1930s, Colebrook and White performed experiments on non-uniform roughness and proposed the following relationship:

$$\frac{1}{\sqrt{\lambda}} = -2\log\left(\frac{k_s}{3.7D} + \frac{2.51}{Re\sqrt{\lambda}}\right)$$ (6.31)

which is applicable to the whole of the turbulent region for commercial pipes, where k_s is Nikuradse's coefficient. Using an effective roughness (k_s/D), determined empirically for each type of pipe, Equation (6.31) may then be solved for λ when Re is known. Finally, for transitional turbulent flow, $2000 < Re < 4000$, the Hazen–Williams equation is used widely in the water industry:

$$V = 0.355 C_{HW} D^{0.63} S^{0.54}$$ (6.32)

where C_{HW} is the Hazen–Williams coefficient. Recommended values for the various coefficients may be found in tables for the design of pipelines.

6.4.2 Examples

Example 6.13. Storm water containment

Statement of problem – A flood storage tank has a volume of 3000 m³. Flood water flows into the tank at a rate $Q_{in} \sim$ Rayleigh(1000) m³/h and drains from the tank through an orifice at a rate $Q_A \sim$ Rayleigh(900) m³/h. Find the probability that the storage tank floods within 3 hours, on the basis that the tank is initially half full and that the time variation can be treated as independent hourly events.

The *failure criterion* is that the net addition of water to the tank exceeds 1500 m³ in 3 hours. The *basic variables* are the inflow rate and the drainage rate. We can use the fact that time variation is divided into independent hour-long blocks. The situation is slightly more complicated than the cases discussed in Section 5.8.2 because, although the flows in individual time blocks are independent, they are linked through the cumulative net volume change in the tank. The *method of solution* is to use Equation (3.61) to estimate the mean and variance of the distribution of the volume of water in the tank, and then determine the probability of failure from these.

Solution. The mean and variance of a Rayleigh (b) variable are $b(\pi/2)^{0.5}$ and $[2 - (\pi/2)]b^2$ respectively. So $E(Q_{in}) = 1253\,\text{m}^3$, $\text{Var}(Q_{in}) = 429,200\,\text{m}^6$, $E(Q_{out}) = 1128\,\text{m}^3$ and $\text{Var}(Q_{out}) = 347,700\,\text{m}^6$. Write $Q_{in}(i)$ and $Q_{out}(i)$ for the inflow and drainage rates, respectively, in hour i. The probability of flooding occurring by the end of the first hour is $P[Q_{in}(1) - Q_{out}(1)] > 1500$; by the end of the second hour it is $P[Q_{in}(1) + Q_{in}(2) - Q_{out}(1) - Q_{out}(2)] > 1500$; and by the end of the third hour it is $P[Q_{in}(1) + Q_{in}(2) + Q_{in}(3) - Q_{out}(1) - Q_{out}(2) - Q_{out}(3)] > 1500$.

The Q terms are independent random variables, so we may use Equation (3.61) to determine the mean and variance of the sum of the inflows and drainage for the 3 hours as:

$$E(V) = 3 \times 1253 - 3 \times 1128 = 375$$

$$\mathrm{Var}(V) = 3 \times 429,200 + 3 \times 347,700 = 2,331,000$$

The approximate probability of failure is given by:

$$\Phi(-375/\sqrt{2,331,000}) = \Phi(-0.2456) = 1 - \Phi(0.2456) \approx 0.4$$

Note that this is only approximate, because the sum of the six Rayleigh distributed variables is not exactly normally distributed. An alternative would be to calculate the exact distribution function of the sum of six Rayleigh variables, or to use the approximate distribution transformation described in Section 5.6.4. However, as the number of variables increases, so the central limit theorem (see Chapter 3) works in favour of the approximation that the resultant distribution is normal.

Example 6.14. Storm water containment

Statement of problem – In an emergency flood operation, $100\,\mathrm{m}^3$ of water is pumped, under constant pressure equivalent to a head of 30 m, from a container through an orifice with area $A = 3.0\,\mathrm{m}^2$, and which has a coefficient of contraction of $C_C \sim N(0.7, 0.1)$ and a coefficient of velocity of $C_V \sim N(0.96, 0.01)$, both of which may be considered to be independent. If the water has to be pumped out within 2 seconds, what is the probability of this not being achieved? The *failure criterion* is that the water is not pumped out of the container in under 2 seconds. The *basic variables* are the coefficients of contraction and velocity. The distribution functions of the basic variables and the failure function is nonlinear, so an appropriate *method of solution* would be a *Level 2 method*.

Solution. Let the volume of water to be pumped be V, and the rate at which it exits the orifice be Q, then the time taken to pump the water is V/Q. The reliability function may be written as:

$$G = 2 - V/Q = 2 - 100/C_V C_C A\sqrt{2gH} = 2 - 33.3/C_V C_C\sqrt{60g} \quad (6.33)$$

or

$$G = 2 - 1.37 C_V^{-1} C_C^{-1} \quad (6.34)$$

which is a nonlinear function of the basic variables. As the basic variables are normal and independent, we will use the Level 2 MVA and FDA approaches.

Table 6.13 Summary of the MVA calculations for Example 6.14

Variable	Mean	Standard deviation	Partial derivative	α_i^2
C_C	0.7	0.1	2.91	0.085
$\underline{C_V}$	0.96	0.01	2.12	0.0004

Thus, with the MVA (Equations 5.14–5.16),

$$\frac{\partial G}{\partial C_C} = 1.37 C_V^{-1} C_C^{-2}$$

$$\frac{\partial G}{\partial C_V} = 1.37 C_C^{-1} C_V^{-2}$$

Now,

$$\mu_G = \mu_R - \mu_s = 2 - 2.04 = -0.04$$

and

$$\sigma_G^2 = \sum_{i=1}^{2} \alpha_i^2 = 0.0854$$

The probability of failure is thus $\Phi(0.04/0.292) = \Phi(0.137) = 0.555$ (see Table 6.13).

With the FDA, it is necessary to transform to normalised variables:

$$z_1 = \frac{C_C - 0.7}{0.1}; z_2 = \frac{C_V - 0.96}{0.01}$$

and the reliability function becomes:

$$G = 2 - 1.37(0.1z_1 + 0.7)^{-1}(0.01z_2 + 0.96)^{-1}$$

or

$$0.096z_1 + 0.007z_2 + 0.001z_1z_2 - 0.013 = 0 \qquad (6.35)$$

Substituting $z_i = \beta\alpha_i$ in the above, following the Hasofer–Lind definitions in Chapter 5 gives

$$\beta = \frac{0.013}{0.096\alpha_1 + 0.007\alpha_2 + 0.001\alpha_1\alpha_2\beta}$$

Table 6.14 Iteration of variables for Example 6.14

	Iteration number			
	Initial value	1	2	3
β	3.0	17.333	−0.135	−0.135
K		0.0873	0.0963	0.0963
α_1	−0.50	−1.000	−0.997	−0.997
α_2	−0.50	0.019	−0.0741	−0.0741

By use of Equations (5.21)–(5.29):

$$\alpha_1 = -\frac{(0.864 + 0.058\beta\alpha_2)}{K}$$

$$\alpha_2 = -\frac{(0.32 + 0.058\beta\alpha_1)}{K}$$

$$\alpha_3 = \frac{0.7}{K}$$

and

$$K = \sqrt{(0.864 + 0.058\beta\alpha_2)^2 + (0.32 + 0.058\beta\alpha_1)^2 + (0.7)^2}$$

The iteration is now initiated by choosing starting values for β, α_1 and α_2, and then calculating new values until convergence is achieved. This process is shown in Table 6.14. As noted in Example 6.9a–d, convergence tends to be faster if positive values are used for α values related to loading variables and a negative sign for those relating to resistance variables. To be contrary, the starting values of the loading variables are set to a negative value.

The probability of failure is then

$$P_f = \Phi(-\beta) = \Phi(0.135) = 0.553$$

This is extremely similar to the value estimated using the less accurate MVA method. In this case, FDA could be used without much loss of accuracy. Note also that the coordinates of the z-design point are negative, indicating that the mean values of the variables lie in the failure region, and hence a probability of failure greater than 0.5 is not unexpected.

Example 6.15. Investigation of the effect of uncertainty in resistance coefficients – based on Y. K. Tung (pers. comm.)

Statement of problem – The Manning, Darcy–Weisbach and Hazen–Williams resistance coefficients are often used in hydraulic calculations and analyses. In general practice, nominal values of these resistance coefficients are used on the basis of experience or published literature. By accounting for the uncertainties inherent in these coefficients, examine the consistency, from a probabilistic view, as they are applied to steady uniform full flow in circular pipes, in particular, riveted steel pipes and cast iron pipes. There is no *failure criterion* in this case. The *basic variables* are the resistance coefficients n, λ and C_{HW}. The question is posed in the form of a research investigation, but probabilistic methods are mentioned.

Solution.   The Manning, Darcy–Weisbach (DW), and Hazen–Williams (HW) equations are among the most frequently used formulae in hydraulic analysis of flow in pipes and channels. Their use requires the specification of the corresponding resistance coefficient, for example, tables in Chow (1959) for the Manning n, the Moody diagram or Colebrook–White formula for the Weisbach λ, and King's handbook (Brater & King, 1976) for C_{HW}. The formulae can be linked nondimensionally as

$$\frac{V}{\sqrt{gRS}} = \sqrt{\frac{8}{\lambda}} = \frac{K_n R^{1/6}}{n\sqrt{g}} = C_{HW}\left(\frac{K_{HW}R^{0.13}S^{0.04}}{\sqrt{g}}\right) \tag{6.36}$$

where V is the velocity, S is the friction or energy slope, A is the flow cross-section area, $R = D/4$ is the hydraulic radius, with D being the pipe diameter, g is gravitational acceleration, K_n is a unit conversion factor (having the dimensionality of g and a value equal to 1 $m^{1/2}/s$ for SI units or 1.486 $ft^{1/3} - m^{1/6}/s$ for Imperial units) (Yen 1992), and K_{HW} is the unit conversion factor for the HW formula, being 0.849 for SI units or 1.318 for Imperial units.

For a given pipe and material, the uncertainty of the resistance coefficient values gives rise to uncertainty in the calculation of the discharge or velocity. Comparison of the Manning, DW and HW formulae can be made in two ways. One is from the user's viewpoint, in which a value of the resistance coefficient is chosen for a specific channel or pipe to compute the flow velocity, discharge or friction slope. In this way the formulae are compared on the basis of the values of the resistance or roughness coefficients given in authoritative references, without considering the compatibility of their ranges and distributions. The other way is to select the range and distribution of the resistance coefficient values of a given formula from which the statistical features of the resistance coefficient of the other formulae are determined for comparing with the suggested values in literature. Values of the Manning n and C_{HW} for different pipe materials can be found directly in standard hydraulics reference books. The Weisbach friction factor λ, on the other hand, may be determined from the Colebrook–White formula.

Due to spatial nonhomogeneity of wall roughness and temporal variations of the flow surface, the values of flow resistance coefficients cannot be assessed with certainty. In natural or man-made open channels, ranges of variation in the Manning n for different types of channel boundary can be found in Chow (1959) and Zipparro and Hasen (1972). For various types of pipe, ranges of the Manning n, pipe wall roughness k_s, and C_{HW} can also be found in literature (Williams & Hazen 1955, Giles 1962, Zipparro & Hasen 1972, Brater & King 1976, Johnson, 1998). As a result, the computed average flow velocity, discharge and frictional loss can be uncertain. By explicitly considering the uncertain nature of the resistance coefficients, the practical implications of uncertainty in resistance coefficients on the computed velocity of steady uniform flow in circular pipes can be determined.

From a user's viewpoint, an engineer can choose any one of the three flow equations as shown in Equation (6.36) to compute the discharge or average flow velocity in a circular pipe for a specified pipe size and design hydraulic gradient. For a specified pipe diameter and friction slope, the probability density function associated with the calculated flow velocity from the three flow formulae can be derived by the transformation-of-variables scheme (see Chapter 3) as:

$$hV(v_n) = gR(n) \times |dn/dv_n|$$
$$hV(v_k) = gR(k_s) \times |dk_s/dv_k| \tag{6.37}$$
$$hV(v_c) = gR(c) \times |dc/dv_c|$$

where $hV(\bullet)$ and $gR(\bullet)$ denote, respectively, the PDFs of a random resistance coefficient and the corresponding flow velocity under consideration; and v_n, v_k and v_c represent the flow velocities computed from the Manning, DW, and HW formulae, respectively. The derivative terms in $|\bullet|$ are the Jacobians, which can be easily obtained for the Manning and HW formulae as

$$\frac{dn}{dv_n} = -\frac{K_n R^{2/3} S^{1/2}}{v_n^2}$$
$$\frac{dC_{HW}}{dv_c} = \frac{1}{K_{HW} R^{0.63} S^{0.54}}$$

and through the chain rule for the DW formula as:

$$\frac{dk_s}{dv_k} = \frac{\partial k_s}{\partial \lambda} \frac{\partial \lambda}{\partial v_k} = -\left(\frac{16gRS}{v_k^2}\right)\left(\frac{2.129D \exp\left(-1.1513/\sqrt{\lambda}\right)}{\lambda^{1.5}}\right)$$

Once the PDFs of flow velocity have been determined, their statistical properties, such as the mean and standard deviation, can be compared to give a more complete indication of the relative performance of the three formulae.

Typical ranges of wall roughness, Manning and HW coefficients appearing in the most popular reference work are listed in Table 6.15.

Table 6.15 Adopted range of resistance coefficients of some selected pipes

Pipe type		Lower bound	Mode	Upper bound
Riveted	Wall Roughness,[1]k_s	0.914 mm	1.829 mm	9.144 mm
steel	Manning's,[2]n	0.013	0.016	0.017
pipes	Hazen-Williams,[3]C_{HW}	80	110	150
Cast	Wall Roughness,[1]k_s	0.122 mm	0.244 mm	0.610 mm
iron	Manning's,[2]n	0.011	0.014	0.016
pipes	Hazen-Williams,[3]C_{HW}	100	110	140

Note: [1]from Giles (1962); [2]from Chow (1959); [3]from Zipparro & Hansen (1972).

For the sake of argument, suppose that each resistance coefficient has a triangular distribution extending over the respective ranges, i.e.,

$$g_R(r) = \begin{cases} \dfrac{2(r-r_l)}{(r_u-r_l)(r_m-r_l)} & \text{for} \quad r_l \leq r \leq r_m \\ \dfrac{2(r_u-r)}{(r_u-r_l)(r_u-r_m)} & \text{for} \quad r_m \leq r \leq r_u \end{cases} \tag{6.41}$$

where R is the random resistance coefficient, which can be Manning's n, C_{HW} or pipe wall roughness k_s; the subscripts l, m and u represent the lower bound, mode, and upper bound of the random variable. The mode adopted here corresponds to the design value or typical value in the literature.

Let us assume that the fluid in the pipe is water with a kinematic viscosity $v = 9.838 \times 10^{-3}$ m^2/s$(1.059 \times 10^{-5}$ ft^2/s) flowing in a circular pipe subject to a friction slope of $S = 0.005$ m/m. The variations in PDF of the resulting flow velocity found by applying the three flow formulae to the two types of pipe are shown in Figure 6.22(a, b) for two different pipe diameters. It is interesting to observe from the figure that the pipe wall roughness k_s-based velocity PDFs in the two different pipe types behave rather differently from the others. For riveted steel pipes, the PDFs of the flow velocity (Figure 6.22a) corresponding to the DW and Manning formulae overlap closely, with the latter formula yielding a slightly more concentrated distribution. Conversely, the PDFs of flow velocity by the DW and Manning formulae in cast-iron pipes (Figure 6.22b) are distinctly separated, while the PDF from the HW formula straddles between them. This gives some indication of the impact of the uncertainty of published resistance coefficient values on the flows in the two pipe types.

As shown in Figure 6.22, the PDF of flow velocity in both types of pipe, as expected, shifts to the right as pipe size increases, indicating an increase in flow velocity with pipe diameter for fixed friction slope. Further, for a given

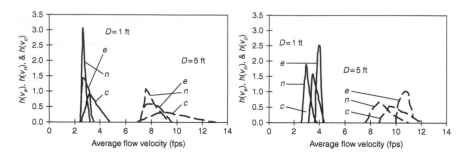

Figure 6.22 Probability density functions for computed flow velocities (in feet per second) using different resistance coefficients: (a) riveted steel pipe; (b) cast-iron pipe. Note: Imperial to metric length unit conversion is 1 foot ~0.3048 m. The labels *e*, *n* and *c* refer to the Darcy–Weisbach, Manning and Hazen–Williams formulae, respectively.

flow formula, the range of variation in the calculated flow velocity also widens with pipe size. Although the range of flow velocity can be calculated in a deterministic way, it cannot provide information with regard to the likelihood of possible velocity variation within the range as a probabilistic analysis could do. Information contained in a PDF can be collapsed into a few pertinent statistical moments. Figure 6.23(a, b) shows the variation of the mean (solid lines) and the coefficient of variation (dotted lines) of n-, k_s-, and C_{HW} -based flow velocities with respect to pipe size. Figure 6.23a clearly shows that for riveted steel pipes the mean and coefficient of variation (C_v = standard deviation/mean) of flow velocity associated with the Manning and DW formulae are similar, whereas those with the HW formula are much higher and wider.

For cast iron pipes (Figure 6.23b), the use of the HW and DW formulae would result in a persistently higher mean flow velocity than that of the

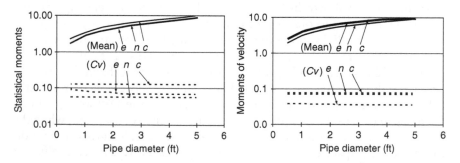

Figure 6.23 Statistical moments of flow velocities (in feet per second) determined by different resistance formulae: (a) riveted steel pipe; (b) cast-iron pipe. The labels *e*, *n* and *c* refer to the Darcy–Weisbach, Manning and Hazen–Williams formulae, respectively.

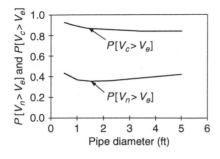 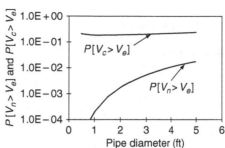

Figure 6.24 Comparison of exceedance probability of flow velocities by different resistance formulae: (a) riveted steel pipe; (b) cast-iron pipe. The labels *e*, *n* and *c* refer to the Darcy–Weisbach, Manning and Hazen–Williams formulae, respectively.

Manning formula. However, the coefficient of variation associated with the flow velocity determined by the DW formula is persistently smaller than that of the two other flow formulae. This is consistent with the plots shown in Figure 6.22.

By adopting the DW formula as the base model, the probability that the *n*- or C_{HW}-based velocity is greater than the k_s-based velocity, that is, $P(V_n \geq V_k)$ and $P(V_c \geq V_k)$, can be calculated and the results are shown in Figures 6.24a,b for the two pipe types.

Although the range of variation of *n*- and k_s-based flow velocities are closer in riveted steel pipes, because the mass of $hV(v_n)$ is more concentrated on the lower velocity range as shown in Figure 6.22a the value of $P(V_n \geq V_k)$ will be slightly less than 50% (see Figure 6.24a). Relative to the HW formula, $hV(v_k)$ lies within the lower half of $hV(v_c)$. This is reflected in Figure 6.24a, which shows a high probability that the HW formula would yield larger flow velocity than the DW formula.

The different ranges of probability distributions shown in Figure 6.22 indicate that the ranges of values of *n*, C_{HW}, and k_s from the literature and given in Table 6.15 are not entirely consistent. Which one is likely to yield the most accurate computed velocity or discharge depends on the range values listed for the particular pipe material. Using the suggested wall roughness for cast-iron pipes in Table 6.15, along with the DW formula, would very likely yield higher flow velocities than would be obtained using the Manning formula (Figure 6.24b). This can be explained by referring to Figure 6.22b, which shows that the great majority of the probability mass in $hV(v_k)$ is on the right-hand side of $hV(v_n)$, with very little overlapping area between the two. If nothing else, this illustrates the uncertainties in predicting the nature of turbulent flow, and the care required when designing drains.

6.5   Maritime and offshore structures

6.5.1   *Introduction*

In this section problems concerning the stability of floating objects and ship collisions with structures are covered. A brief summary of the stability of floating objects is given first, followed by introductory material on ship damage to offshore structures.

At its simplest, the stability of floating objects reduces to an analysis of the relative magnitudes and centres of action of the gravitational force and the buoyancy force acting on an object. The buoyancy force is described by Achimedes' principle. This is illustrated in Figure 6.25.

Now,

$$F_{p_1} = \rho gy\,\pi R^2$$
$$F_{p_2} = \rho g(y+h)\,\pi R^2$$
$$\text{Upthrust} = F_{p_2} - F_{p_1} = \rho gh\,\pi R^2$$

i.e., Upthrust = weight of liquid displacement

For floating objects, such as ships, we need to consider the combined effect of the buoyancy and weight forces. Figure 6.26 shows the force diagram for a stable body, where W is weight and U is upthrust. If the body is displaced slightly to one side the upthrust no longer acts through the centre of gravity G. The stability can be analysed by looking at the position of the metacentre M. The metacentre is the point of intersection of a vertical line

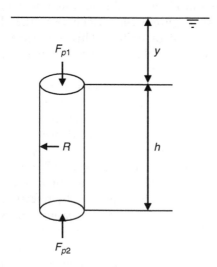

Figure 6.25 Forces acting on a floating body – in this case a cylinder of radius R.

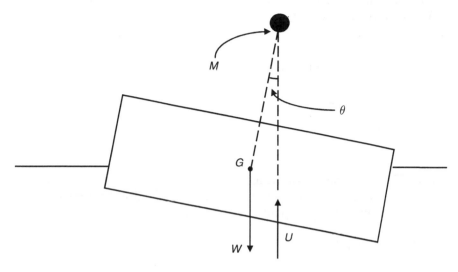

Figure 6.26 Illustration of forces acting on a floating body and the metacentre, *M*.

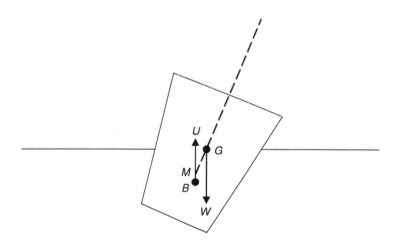

Figure 6.27 Unstable body, with *GM* < 0 (i.e., pointing downwards), and the upthrust and weight creating a couple acting to overturn the body.

up from the line of action of the upthrust *U* and the centre line of the body. The upthrust acts through the centre of buoyancy, B, which is the centre of gravity of the displaced liquid. In this case the restoring couple is positive as *M* is above *G*. *GM* is known as the metacentric height. The couple acting on the body is *GMW*sin(θ). Figure 6.27 shows the force diagram for an unstable body.

The restoring couple is negative and the metacentre is below G. Thus, GM is a measure of stability:

$GM > 0 \Rightarrow$ Stable

$GM < 0 \Rightarrow$ Unstable

It can be shown that

$$BM = \frac{I}{V}$$

where BM is the distance of the metacentre above the centre of buoyancy, I is the waterline second moment of area, and V is the displaced volume.

The following relationship may be seen to hold:

$$BM = BG + GM$$

where BG is the distance between the centre of buoyancy and the centre of gravity. It can also be shown that the period of oscillation T is given by

$$T = 2\pi \sqrt{\frac{I_R}{W \times GM}} \tag{6.38}$$

where I_R is the moment of inertia about the roll axis.

The following summarises key elements of the Health and Safety Executive report on 'Comparative Evaluation of Minimum Structures and Jackets'. In recent years there has been a trend to use minimum facility platforms (MFPs) to 'fast track' the development of marginal oil and gas fields in water depths of up to 60 m. Similar structures are also being considered for deepwater wind farms, for which monopile structures are insufficient. MFPs are finding favour primarily due to the relatively low fabrication and installation costs. In contrast to conventional jackets, MFPs have a slender construction with low stiffness and a low level of redundancy. The HSE was concerned to understand the potential sensitivities of such structures to damage caused during design, construction and operation. A project was set up, described in HSE (2002), to evaluate the reliability of three MFPs, and to compare the results against those from a conventional four-pile jacket under extreme loading conditions. The structures considered were: (i) a three-pile monotower; (ii) a Vierendeel tower; (iii) a braced caisson; and (iv) a four-pile jacket. The three MFPs are shown in Figure 6.28 and an example of a four-pile jacket is shown in Figure 6.29. Detailed numerical simulations were performed to study the performance of these structures against collision from a supply vessel. Following the impact, post-impact pushover analyses were performed to determine the reduction in pushover capacity

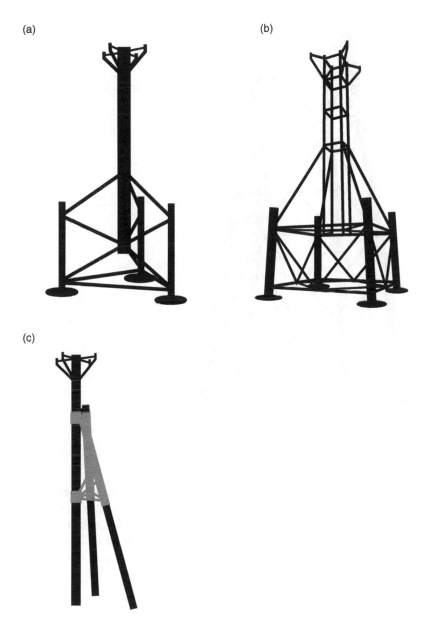

(a)

(b)

(c)

Figure 6.28 The three MFP structures: (a) three-pile monotower; (b) Vierendeel
tower; (c) braced caisson.

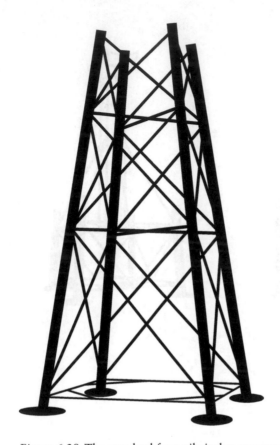

Figure 6.29 The standard four-pile jacket structure.

resulting from the collision. For each structure, analyses were performed for a range of vessel sizes (500–3500 tonne mass) and impact velocities up to 2.5 m/s, which covered the expected range of values for operations in the North Sea fields. The report concludes that MFPs can be made to be as reliable as conventional jackets by:

- Designing for ship impact to mitigate the risk of damage by considering the dynamic interaction between vessel and structure;
- Designing critical welds for fatigue lives > 10 times the service life;
- Using a reliability-based approach during design.

Environmental data and soil properties were taken from the Davy field in the North Sea. The key environmental parameters are summarised in Table 6.16.

Table 6.16 Environmental parameters used for design

Water depth including storm surge	36.2 m
100-year return wave height	16.4 m
Period of the 100-year wave	12.6 s
Associated current speed at the surface	0.96 m/s
Associated wind speed (1 hour mean @ 10 m above LAT)	32.2 m/s

Conventional North Sea practice considers ship collision from a 2500 tonne vessel at 2 m/s. The project team found that this was seen to totally govern the section dimensions of most of the structures, which made the other loading conditions insignificant.

For reliability analysis, the structure was modelled as a single component with its resistance representing the ultimate strength of the structure under the pushover condition. The reduction in pushover capacity due to ship impact damage was modelled using a reduction factor $f(M, v)$, which is a function of the ship mass and velocity of impact.

The reliability function G for collapse of the structure under extreme environmental loading following ship impact is expressed as

$$G = X_{model} \times R_{init}[1 - f(M, v)] - X_{hydro}(aH^b) \times F_{hydro} - X_{wind} \times F_{wind} \quad (6.39)$$

where:

X_{model} = random factor for uncertainty in ship impact and pushover capacity

X_{hydro} = random factor for uncertainty in base shear calculations

F_{hydro} = base shear due to associated hydrodynamic loading on the deck (random)

H = annual maximum wave height (random)

X_{wind} = random factor for uncertainty in wind force calculations

F_{wind} = base shear due to associated wind loading on the deck (random)

R_{init} = ultimate capacity of the structure in terms of base shear at collapse

$f(M, v)$ = function to account for degradation of system strength due to ship impact (see below)

a, b = structure-dependent parameters, fitted from analysis results, to relate base shear wave height.

The function $f(M, v)$ depends on the mass M and velocity v of the ship. The results of the analyses suggested that $f(M, v) = 0$ for all structures except the three-pile monotower. The monotower structure failed during the ship impact for certain combinations of mass and velocity. The reliability function during ship impact for this structure is expressed as

$$G = (411.5v^2 - 3971v + 9560) - M \qquad (6.40)$$

The term in brackets represents the capacity of the structure against ship impact and was obtained by fitting a function to those values of mass and velocity that resulted in the failure of the structure during impact.

The vessel sizes that could visit a structure were modelled using a uniform (rectangular) distribution between 500 and 3500 tonnes. (This does not include the added mass which was taken into account during analyses.) For a given vessel, the uncertainty in its mass was modelled using a normal distribution with a coefficient of variation of 0.15. The velocity of impact was taken to be exponentially distributed with a mean of 0.3 m/s and a standard deviation of 0.3 m/s.

6.5.2 Examples

A typical deterministic problem might be something like the following. A pontoon of density 600 kg/m³, length 4 m, width 2 m and height 0.8 m floats in water (similar to those in Figure 6.30). Calculate the period of oscillation about its long axis. (The moment of inertia about this axis is $Mh^2/12$, where M is the mass and h is the height.)

Solution.   We have $BG + GM = \frac{I}{V}$ and from Equation (6.39) $T = 2\pi\sqrt{\dfrac{I_R}{W(GM)}}$

Figure 6.30 Floating pontoons on the Bedford River, UK (courtesy of Dr Les Hamill).

Now,

$$I = \frac{1}{12} \times 4 \times 2^3 = 2.67\,\text{m}^4$$

and $G = 0.4\,\text{m}$ above base.

The density of water is $1000\,\text{kg/m}^3$, so the pontoon sinks to a depth of $0.6 \times 0.8 = 0.48\,\text{m}$

$\therefore\qquad B = 0.24$ above base

$\therefore\qquad BG = 0.16\,\text{m}$

The submerged volume is $V = 0.48 \times 4 \times 2 = 3.84\,\text{m}^3$

$$\therefore GM = 2.67/3.84 - 0.16 = 0.535\text{m}$$

GM is positive so the pontoon is stable. Finally,

$$I_R = \frac{1}{12} M\,0.8^2 = 0.0533 M\,\text{kgm}^2$$

and

$$T = 2\pi \sqrt{\frac{0.0533M}{M \times 9.81 \times 0.535}} = 0.633\ \text{s}$$

which is independent of the mass.

Example 6.16. Stability of pontoons

Statement of problem – Now suppose in the example above that the pontoon is not homogeneous (which is very likely with a solid timber block, but a certainty with a hollow construction). Its mass remains unchanged, but there will be uncertainty about its centre of gravity and its moment of inertia. Suppose $G \sim U(0.1, 0.7)$ and $I \sim U(0.20V, 0.80V)$. Determine whether the pontoon might be unstable and, if so, what the probability of instability is.

The *failure criterion* becomes clear if the stability of the pontoon is cast as a reliability problem. The pontoon will be unstable if $GM < 0$. Thus, GM may be treated as the reliability function, with I/V being the 'strength' and BG the 'load'. The pontoon 'fails' if $GM < 0$. The *basic variables* are I and G, and these are described by probability density functions; therefore, a *Level 2 method of solution* will be appropriate. In fact, as will be seen, the density function of the reliability function and the probability of failure can be determined directly.

Solution. The reliability function may be written as

$$GM = I/V - BG = I/V - (G - B) = I/V + (B - G)$$

where G and B are the distances of the centre of gravity and buoyancy, respectively, above the base of the pontoon. Now, $I \sim U(0.20V, 0.80V)$ so $I/V \sim U(0.20, 0.80)$. Also, $G \sim U(0.1, 0.7)$, so $B - G \sim U(-0.46, 0.14)$. The probability density of GM is therefore described by the sum of two uniformly distributed variables (I/V and $B - G$), which have the same range of 0.6. The resulting density is symmetric and triangular over the range (1.91, 3.11) (see Chapter 3). As the area under the density function must be unity, the formula for the area of a triangle the peak of the distribution is found to be 5/3 at $GM = 0.34$. Figure 6.31 shows the density function of GM. From basic geometry, the slope of the straight-line segment of the left-hand side of the triangle is $(5/3)/0.6 = 2.78$. Furthermore, when $GM = -0.26, f_{GM} = 0$, so the equation of the straight-line segment is $f_{GM} = 2.78GM + 0.723$. The probability of failure P_f is the area under this straight line to the left of $GM = 0$, that is, the area of the shaded triangle. This can be calculated directly as follows. The area of the shaded triangle $= 1/2 \times$ base $\times$ height $= 0.5 \times 0.26 \times 0.723 = 0.094$. Thus, the probability that the pontoon will be unstable is ~ 0.09.

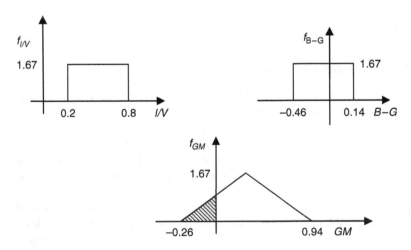

Figure 6.31 Density functions of I/V (top left), $B-G$ (top right) and GM (bottom). The bottom panel also shows the area under the density function corresponding to the failure region $GM < 0$.

Example 6.17. Oscillations of a floating body

Statement of problem – From the theory above, it would seem that the meta-centric height should be made as large as possible to ensure vessel stability. However, this is not always possible or desirable because, as GM increases, so the period of roll decreases. On a passenger ship, human comfort is a major consideration in addition to safety, and a rapidly rocking ship would not be welcomed. As cargo and passengers move around the vessel, both the centre of gravity and the moment of inertia will change. For simplicity, con-sider the centre of gravity to be fixed, with $GM=10$ m and $W=100,000$ kg. The problem is to find the probability that the period of oscillation is less than a critical value T_c, thereby causing discomfort to the passengers. T_c is specified to be 7 seconds.

The *failure criterion* in this case is that the period of roll is less than some predetermined value T_c. The vessel 'fails' if $T < T_c$. The *basic variables* are I_R and T_c. For this problem $T_c = 7$ s and the density of I_R, f_{IR}, is a $U(1,000,000, 1,800,000)$ distribution. As the basic variables are described by probability density functions, a *Level 2 method of solution* will be appropriate.

Solution. First, the reliability function, $G(T, T_c)$, is written as

$$G(T, T_c) = T_c - T = T_c - 2\pi\sqrt{\frac{I_R}{W \times GM}} = T_c - \sqrt{\frac{4\pi^2 I_R}{W \times GM}}$$

Now, from Example 3.5, the density of the period of roll f_T is found to be:

$$f_T = \frac{W \times GM}{4\pi^2} T f_{I_R}\left(\frac{(W \times GM)T^2}{4\pi^2}\right)$$

$$= 25,330T\frac{1}{800,000}$$

$$= 0.0317T$$

The limits on this distribution are found by substituting the limits on I_R into the expression for T, yielding $6.28 \leq T \leq 8.43$. The probability of failure P_f is the trapezoidal area under this straight line to the left of $T = 7$ s (see Figure 6.32).

This can be calculated directly as $0.5 \times (7 - 6.28) \times (0.199 + 0.222) = 0.15$. Thus, the probability of failure is 0.15, so that the chances that the 'Queen of the Seas' will turn into the 'Vomiting Venus' are 15%.

Example 6.18. Environmental loading on an offshore structure – factor of safety

Statement of problem – Using $\Gamma_{model} = 1.2, \Gamma_{hydro} = 1.1, \Gamma_{wind} = 1.15$ and $F_{hydro} = \alpha F_{wind}$, where α is a constant determined from other calculations,

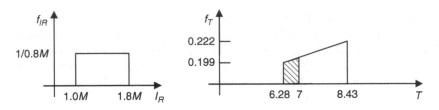

Figure 6.32 Calculation of the probability of failure for Example 6.17.

and the fact that $f(M, v) = 0$, recast Equation (6.40) in terms of the failure function and partial safety factors and determine the global factor of safety. The *failure criterion* will be in the form of 'strength' minus 'loading' and the criteria for failure will be if the reliability function becomes negative. The *basic variables* are the capacity of the structure and the combination of the wind and wave forces. Probability density functions are not provided, rather partial safety factors are specified, and therefore a *Level 1 method of solution* will be appropriate.

Solution. Equation (6.40) becomes

$$G = R - S = \frac{R_{\text{init}}}{\Gamma_{\text{model}}} - \Gamma_{\text{hydro}} \times F_{\text{hydro}} - \Gamma_{\text{wind}} \times F_{\text{wind}} \tag{6.41}$$

where the Γ terms are the performance and partial safety factors and $f(M, v) = 0$. Setting $G = 0$ in Equation (6.42) and multiplying through by Γ_{model} gives

$$R_{\text{init}} = (\Gamma_{\text{hydro}} \times F_{\text{hydro}} + \Gamma_{\text{wind}} \times F_{\text{wind}}) \times \Gamma_{\text{model}}$$

Substituting in the given values for the safety factors, Equation (6.42) becomes:

$$R_{\text{init}} = (1.1 \times \alpha \times F_{\text{wind}} + 1.15 \times F_{\text{wind}}) \times 1.2 = (1.32\alpha + 1.38) \times F_{\text{wind}}$$

So the global factor of safety is $(1.32\alpha + 1.38)$.

Example 6.19. Environmental loading on an offshore structure – reliability index

Statement of problem – Given that $R_{\text{init}} \sim N(1500, 200)$ and $F_{\text{hydro}} \sim N(800, 300)$ – (where these composite strengths and forces are assumed to have been calculated using structural codes and Morison's equation for wave loading, for example) calculate the reliability index and the probability of failure. Wind loading may be neglected. The *failure criterion* will be that the hydrodynamic loading exceeds the structure's capacity. The *basic variables* are the capacity of the structure and the hydrodynamic forces. Probability density functions are provided, so a *Level 2 method of solution* will be appropriate.

Solution. Starting with Equation (6.40) and setting $F_{wind} = 0$, the reliability function G may be written as

$$G = R - S = R_{init} - F_{hydro}$$

Standard results for normal distributions give

$$\mu_G = \mu_R - \mu_S$$

and

$$\sigma_G^2 = \sigma_R^2 + \sigma_S^2$$

Therefore, $\mu_G = 1500 - 800$ and $\sigma_G^2 = 200^2 + 300^2 = 130,000$, so $\sigma_G = 360.6$.
The reliability index is $\beta = \dfrac{\mu_G}{\sigma_G} = \dfrac{700}{360.6} = 1.94$
The probability of failure is thus

$$P_F = \Phi(-\beta) = 1 - \Phi(\beta)$$

We find $\Phi(\beta)$ from the statistical tables for the normal distribution as 0.9738, so that the probability of failure is $1 - 0.9738 = 0.0262$, or about 1 in 40.

Example 6.20. Environmental loading on an offshore structure – specification of strength

Statement of problem – Given that R_{init} has a variance of 400 due to variations in the quality of the steel, and F_{hydro} is N(800, 300), and that wind forces may be neglected, find the mean strength (or resistance to shear) that gives a probability of failure of 0.01. The *failure criterion* will be that the hydrodynamic loading exceeds the structure's capacity. The *basic variables* are the capacity of the structure and the hydrodynamic forces. Probability density functions are specified, so a *Level 2 method of solution* will be appropriate.

Solution. In this example, the problem is cast in the reverse sense to many reliability calculations. Given the density functions of the loading variable and the variance of the strength, it is required to find the mean strength that will give a specified reliability. This problem has one equation (the definition of the reliability index) and one unknown.

A probability of failure of 0.01 corresponds to a value of $\beta = 2.33$ (see Table D in Appendix A).

Now, $\beta = \dfrac{\mu_G}{\sigma_G}$, so we can work out μ_R as follows.

First, $\sigma_G^2 = 400^2 + 300^2 = 250,000$, so $\sigma_G = 500$. Therefore $\mu_G = 2.33 \times 500 = 1165$, and so $\mu_R = \mu_G + \mu_S = 1165 + 800 = 1965$.

Example 6.21. Ship impact

Statement of problem – In the report the consultants used a uniform (rectangular) distribution for the ship mass and an exponential distribution for the ship speed. Using distributions that are not normal (Gaussian) makes the calculations more complicated (but still possible). For this example, both distributions are taken to be normal, with $M = N(4500, 250)$ and $v = N(1.0, 0.1)$. (NB. The choice of normal distributions in this case can lead to non-physical situations, as there is a very small, but not zero, probability of a ship with negative mass and negative speed. This is justified on the basis that interest is in the upper tails of the distribution where both masses and speeds are positive. In practice the choice of distributions that are used to model the physical problem should have suitable properties that reflect reality.) What is the reliability index and probability of failure, assuming the mass and velocities follow the normal distributions above?

The *failure criterion* will be that the ship impact loading exceeds the structure's capacity. The *basic variables* are the mass and speed of the ship. Probability density functions are given, so a *Level 2 method of solution* will be appropriate.

Solution. The mean value approximation is used in this example. For this, the partial derivatives of G with respect to the basic variables are needed. Now, G is given by Equation (6.41) as

$$G = (411.5v^2 - 3971v + 9560) - M$$

The partial derivatives are:

$$\partial G / \partial v = 823v - 3971$$
$$\partial G / \partial M = -1$$

The derivatives at the mean values of the variables are:

$$\partial G / \partial v = 823v - 3971 = 823 \times 1.0 - 3971 = -3148$$
$$\partial G / \partial M = -1$$

Table 6.17 Summary of MVA calculations for Example 6.22

Variable	Mean	Standard deviation	Partial derivative	α_i^2
v	0.5	0.1	−3,148	99,099
M	4,500	250	−1	62,500

Table 6.17 summarises the results of the calculations. The influence factors, α_i^2, suggest that the speed of the ship is rather more important than its mass in determining the failure of the structure, in the ratio of approximately 3:2.

Thus, $\mu_G = [411.5 \times 1.0^2 - 3971 \times 1.0 + 9560] - 4500 = 1500.5$

and

$$\sigma_G^2 = \sum_{1=1}^{2} \alpha_i^2 = 161,599$$

So, $\sigma_G = 402$

The reliability index is

$$\beta = \frac{\mu_G}{\sigma_G} = 1500/402 = 3.73$$

and the probability of failure is $P_f = \Phi(-\beta) = 1 - \Phi(\beta) = 1 - 0.9989 = 0.0011$, or about 1 in 900.

6.6 Coastal protection

6.6.1 Introduction

Shoreline cliffs may be categorised as either active or inactive. An active cliff undergoes erosion as incident waves break against the base and expend their energy. An inactive cliff has a beach at its base that lessens the erosive activity of the waves. Wave erosion on active cliffs carves notches, caves and arches in the cliff rock. When the notch, cave or arch grows sufficiently, the weight of the overhanging rock can no longer be supported and collapses, leaving debris at the base of the cliff or, in the case of an arch, a column of rock, called a sea stack, standing in the water. Figures 6.33 and 6.34 show the results of this type of process.

Particular problems can occur when coast protection is installed to protect conurbations. The village of Happisburgh is a notable example

Figure 6.33 The undefended eroding cliffs at Hunstanton, North Norfolk, UK. Note the large rock debris at the base of the cliff, the result of earlier cliff falls, and the wide sandy beach.

Figure 6.34 Stacks and arches near Vik, Iceland.

Figure 6.35 Rapidly eroding cliffs at Happisburgh, North Norfolk, UK. In the fore-
ground the remains of a timber coast protection can be seen. The centre
of Happisburgh village can be seen at the top of the photograph and
the Happisburgh lighthouse towards the top left. The photograph was
taken in summer 1998. The row of houses at the top of the cliffs and
the three houses lying closest to the sea along the road running diago-
nally from bottom left to middle right have all now been lost to erosion
(Photograph courtesy of Mike Page).

in the UK, but by no means the only one. Here, timber defences were
installed some years ago to defend the base of the cliffs against wave
action (Figure 6.35). Now, at the end of the design life of the defences,
the approach to coastal defence has changed. Societal perception of the risk
is such that the benefit arising from constructing new defences is deemed
insufficiently large to attract government funding. The result is that the cliff
is now eroding at an average of several metres a year, and the owners of
the properties find themselves in a position where they can neither sell in
order to release funds that would enable them to relocate, nor construct
new defences to slow the erosion.

In this section, the development of a probabilistic method for estimating
the risk due to coastal erosion is described. This is included as an example
of how methods are developed to give robust information to engineers and
managers when there may be insufficient measurements to define probability
distributions or even safety factors.

6.6.2   Examples

Example 6.22. Risk assessment of coastal erosion

Statement of problem – In 2005, the UK Department of Environment, Food and Rural Affairs funded a project 'Risk Assessment of Coastal Erosion' (RACE). The aim of this project was to develop, test and disseminate a probabilistic method for assessing the risk of coastal erosion. The method had to be supported by data and information from local monitoring programmes and risk-based inspections, and also be compatible with the existing methods used for flood risk assessment. In this context, decisions have to be made that have practical consequences. In many cases, the luxury of waiting many years to compile detailed measurements was not available, information was limited, and assessment of the hazards and risks was reliant upon local knowledge and engineering judgement.

The *failure criterion* is that erosion over a specified time exceeds a value at which significant assets are lost. The *basic variables* are the capacity of the coastal defence to withstand wave action, and the capacity of the undefended cliff to resist erosion and the hydrodynamic forces. Not only is there minimal information, but there is also the possibility of a variety of different types and quality of data, so a *hybrid method of solution* is probably required.

Solution.   The methodology is based on the source–pathway–receptor (S-P-R) risk model, DETR (2000), together with the concept of using a tiered approach to evaluating the hazards and responses along the lines developed by Meadowcroft et al. (1996a, b) and Environment Agency (1996), and implemented by Hall et al. (2003) for flood risk. Thus, the various sources of the erosive forces and how they propagate to their point of impact are determined before assessing the magnitude of the effect on receptors.

The framework allows a range of probabilistic analyses to be used, so that the method can be tailored to the level of detail required, ensuring that proportionate effort is applied. Figure 6.36 illustrates the general approach

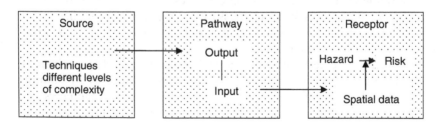

Figure 6.36 Illustration of the Source–Pathway–Receptor risk model.

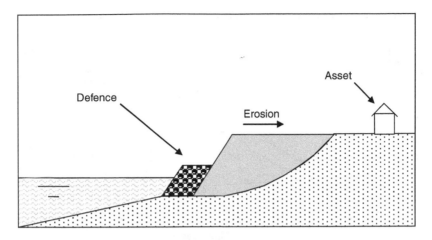

Figure 6.37 Components of the risk-based cliff erosion methodology.

that has been adopted. The source data are the description of the cliff instability and erosion process, that is, the mechanisms and rate at which it might occur (definition of the hazard). The pathway stage represents the response of the defence (if any) and the cliff to the imposed hydraulic forces, that is, coastal protection and slope stability measures that slow erosion and lead to a modification of the hazard. The receptor stage comprises the combination of hazard and consequence to produce the probabilistic risk assessment. This is illustrated in Figure 6.37.

The general idea is to define, by means of a probabilistic description of failure and erosion, a probability of erosion for a given distance or time, which considers the interaction of both elements: the natural process of cliff erosion and the presence or not of a coastal structure. The strategy is, first of all, to define a time window of interest and predict a time history of future erosion, with uncertainty bands. Second, where there is a coastal structure, then a probability of failure over the same time period is generated, again with uncertainty bands. With these two graphs it is possible to generate probabilities of erosion to a fixed asset or, with many points along the cliff edge, the spatial distribution of erosion at a given time in the future.

Risk assessment of cliff erosion

With regard to cliff erosion there is uncertainty over the time as to when the cliff will erode, by how much and whether the erosion is instantaneous or gradual. With regard to resistance to cliff instability, there are uncertainties relating to when, or if, any coastal protection may fail, whether it would be reinstated, and any impacts of sea-level rise. The key components required to determine the erosion versus time plot are the cliff instability

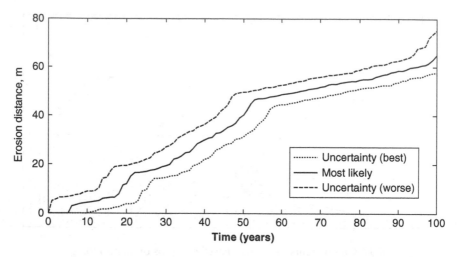

Figure 6.38 Erosion–time graphs, showing best estimate and bounds of uncertainty. These may be obtained from numerical models of erosion, extrapolation of historical data or engineering judgement.

and erosion processes. These curves might be generated on the basis of detailed cliff erosion models, on extrapolation of historical erosion rates, or local knowledge and engineering judgement. At a well-monitored site it may be possible to use a quantitative model to predict cliff erosion rates (see e.g., Lee et al. 2001). Conversely, in a situation where there is little or no recorded information, local knowledge and engineering judgement are necessary. Irrespective of the simplicity or complexity of the techniques, it is a requirement that the same basic output is provided: a timeline to cliff erosion and a timeline to defence failure. The principal difference between the outputs is the accuracy and the level of confidence.

Figure 6.38 illustrates a simple example of the required input, with the black solid line indicating the best assessment, and the degree of uncertainty shown by the dotted and the dashed lines.

So, in the above example, the cliffs are expected to erode inland by an average distance of 64 m over the next 100 years, although there is potential that the cliffs could erode by as little as 58 m, or as much as 76 m.

Coastal defence assessment

Similarly, the probability of defence failure can be determined. In all cases, there are two components:

- a general deterioration over time due to general wear and tear, which at some point in the future will render the defence ineffective;

- destruction of the defence arising from circumstances exceeding the design conditions; for example, damage due to major storms, or undermining due to falling beach levels.

Both of these are variable but in different ways. With regard to deterioration, there is uncertainty over the time at which the defence will become ineffective, and indeed if this would be instantaneous or gradual. Failures resulting from environmental conditions can be determined from an annual probability of exceedance. Factors including climate change and alongshore interactions can also be incorporated into any of the techniques using appropriate allowances.

Figure 6.39 illustrates the input required by the methodology, with the solid black line indicating the best assessment, and the degree of uncertainty shown by the dotted and the dashed lines, which are considered to approximate the 5% and 95% confidence limits.

In this example, it has been determined that the structure is most likely to be effective for another 55 years, but it might collapse after 39 years, or it could last 65 years. During the period leading up to that, it has been estimated that there is a ~0.6% chance of storm conditions exceeding design conditions year-on-year and leading to its failure.

In other words, under the 'best case' the graph is saying that the defence will definitely have failed by year 65, but recognises that there is a small chance that this failure could actually happen this year, next year, or at any point beyond that.

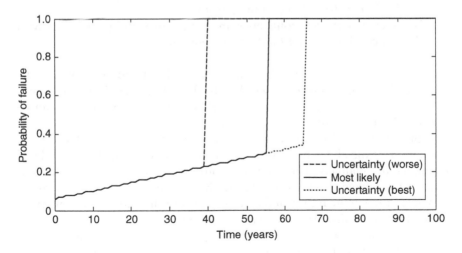

Figure 6.39 Probability of failure–time graph.

Range of techniques for determining recession and probability of failure

From Figures 6.38 and 6.39 it is possible to work out, for any year in the future, the probability of the defence failing and the consequent erosion from that point onwards. For a fixed year in the future all these probabilities can be combined to estimate the probability density of erosion distance. Alternatively, for a fixed position inland of the initial cliff position, one may determine the probability of the erosion extending to that point as a function of time. As there are upper and lower bounds to the erosion and probability of defence failure, the calculations can be repeated using various combinations of the erosion curves and probability curves. Furthermore, when a defence ceases to be effective, a period of accelerated erosion is often found to occur, as the once defended cliff 'catches up' with the eroded cliff either side of it. These variations were included in a spreadsheet calculator developed for the project.

Definition of scenarios

The input values, erosion distances and probabilities of failure are defined with three different possibilities: most likely, best case and worse case scenarios. Thus, there are nine different combinations that can be plotted. However, given that practitioners required the most likely, best and worst case scenarios, only these three combinations are calculated. The combinations are specified as follows:

- Worst scenario: the 95% confidence limit of the probability of failure with the 95% confidence limit on the erosion curve.
- Intermediate: the 95% confidence limit of the probability of failure with the 50% confidence limit on the erosion curve.
- Best scenario: the 5% confidence limit of the probability of failure with the 5% confidence limit on the erosion curve.

To account for 'catch up' the user may specify a time over which the arrested cliff erodes to where it would have been had there been no defence in place. Further information on the methodology and spreadsheet can be found in DEFRA (2006) and Pedrozo-Acuña et al. (2008).

Figures 6.40–6.42 illustrate the type of plots that can be produced from the output of the above procedure, and which are of particular use to local engineers and planners. These include: contour maps of the probabilities of erosion for a fixed time; maps showing the probability of asset loss, again at a given time in the future; and contour maps of the expected evolution of the cliff line over time, together with uncertainty bands.

By producing a sequence of plots like the one shown in Figure 6.40b it is possible to build up contour plots of zones of equiprobable erosion at a fixed time, as shown in Figure 6.40a. Similarly, by repeating calculations for Figure 6.41b at different points, it is possible to build up a map showing

(a)

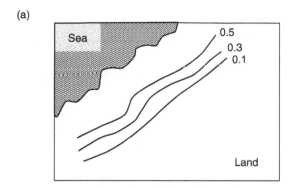

Figure 6.40a Typical mapping output – zones of equiprobable erosion at a fixed time.

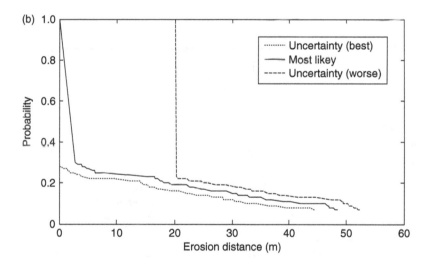

Figure 6.40b Typical spreadsheet output – probability of erosion after 60 years.

the probability of loss due to erosion at a particular year in the future, as illustrated in Figure 6.41a. Alternatively, from a sequence of plots such as the one in Figure 6.40a it is possible to construct 'expected' bands of cliff-line evolution for different years by selecting contours corresponding to particular quantiles (e.g., 5% and 95%).

The different types of model that are available for forecasting cliff and defence response are described in detail in DEFRA (2006). In practical engineering applications other considerations, apart from purely scientific ones, come into play. Thus, the quantity and quality of data can be a very

(a)

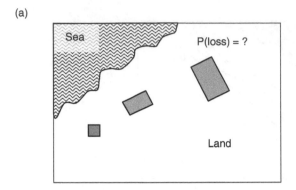

Figure 6.41a Typical mapping output – probability of losing a fixed asset at a given time.

(b)

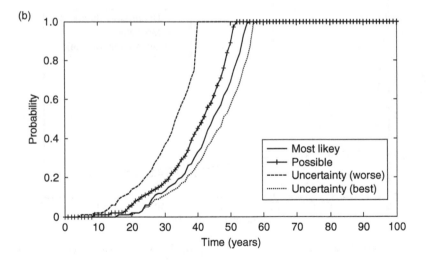

Figure 6.41b Typical spreadsheet output – probability of erosion to a fixed point.

important factor in determining the type of model used. In fact, the quality of the data is one of the main sources of uncertainty in modelling. Even if the predictive model is perfect, if inaccurate data are used as input, the forecast will be affected. In situations where the available information is of poor quality, a model that gives qualitative forecasts may actually be of more practical assistance in informing a decision than an inaccurate forecast produced by a sophisticated model.

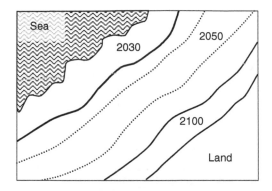

Figure 6.42 Typical mapping output – expected bands of cliff-line evolution incorporating bands of uncertainty defined by quantiles of the distributions determined in Figure 6.38a.

Example 6.23. Monte Carlo simulation of cliff erosion

Statement of problem – An awareness of the likely cliff position at some future date is crucial for coastal planning and management. Reliable predictions of coastal recession rates are needed for effective land-use planning. Investigate methods that might be employed to perform cliff-recession predictions.

No *failure criterion* is defined explicitly, but one could imagine that a possible criterion might be that the cliff top should remain a specified distance seaward of a particular asset. The *basic variables* are the cliff position and the wave, water level and beach level variables. After some preparatory work it is likely that the *method of solution* will be based on a variety of approaches.

Solution. The papers by Lee et al. (2001) and Hall et al. (2002) give a description of a variety of methods tested for predicting cliff erosion. These range from a Monte Carlo type prediction of individual cliff falls, in which the frequency and size of cliff falls are taken as random variables, to the use of historical records. In their Monte Carlo modelling, cliff recession is considered to proceed as a series of discrete cliff-fall events. A discrete model of the cliff recession X_t during duration t is specified as:

$$X_t = \sum_{i=1}^{N} C_i$$

where N is a random variable representing the number of cliff falls in duration t and C_i is the magnitude of the ith fall. The model is defined

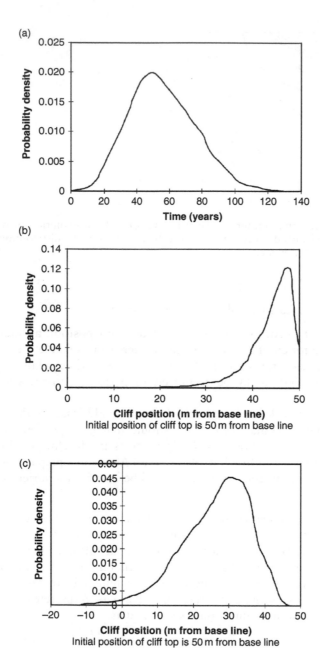

Figure 6.43 Sample results from Lee et al. (2001) showing (a) predicted time to recess by 29 m; (b) predicted cliff position at Year 10; (c) predicted cliff position at Year 50 (Reprinted from *Geomorphology*, Vol. 40, Lee, Hall & Meadowcroft, Coastal cliff recession: the use of probabilistic prediction methods, Copyright [2001], with permission from Elsevier).

by two distributions, one describing the timing of cliff falls and the other their magnitude. The arrival of storms is assumed to follow a Poisson distribution. After a sufficient number of storms a cliff-recession event occurs. The time between successive events can thus be described by a Gamma distribution. The magnitude of the cliff falls is described by a log-normal distribution, which is chosen on the basis of laboratory experiment results. The Monte Carlo model thus requires the specification of four parameters: the shape and scaling parameters of the gamma distribution, and the mean and variance of the log-normal distribution. Cliff-fall events and magnitudes are then generated using a random-number generator. A single realisation consists of a simulation of cliff falls over a specified period of time. This simulation is then repeated many times over to build up a picture of the distributions of cliff recession. Figure 6.43 shows some results from this model.

6.7 Coastal morphology

6.7.1 *Introduction*

Prediction of beach evolution over scales of tens of kilometres is very important for coastal management and scheme design (e.g., Cowell et al. 2006). It is difficult because of the high complexity of the coastal system (e.g., De Vriend 2003, Southgate et al. 2003). One mathematical tool that is used widely for predicting beach evolution is the 'one-line' model. This is an idealised model of beach dynamics proposed by the French engineer Pelnard-Considère (1956). It is based on the assumption that the shape of the beach profile is in equilibrium, that the shoreline evolves as a result of spatial gradients of the alongshore sediment transport, and the principle of the continuity of sediment. Cross-shore transport and other complex mechanisms are parameterised in a simplified manner. The model predicts the position of a single beach level contour, hence the sobriquet 'one-line'.

Consider the element of beach between boundaries 1 and 2 in Figure 6.44b, shown in sectional elevation in Figure 6.44c. Applying the continuity equation, in a time interval δt the change in the volume of sediment in the element is equal to the volume entering less the volume leaving, yielding

$$\frac{\partial Q}{\partial x} = \frac{\partial A}{\partial t} \tag{6.42}$$

where Q is the volumetric a longshore transport rate and A is the cross-sectional area of the element. For an equilibrium profile, any change in area must result in a horizontal movement of the profile δy, given by $\delta A = d_c \delta y$, where d_c is known as the 'depth of closure' and is the point at which it is

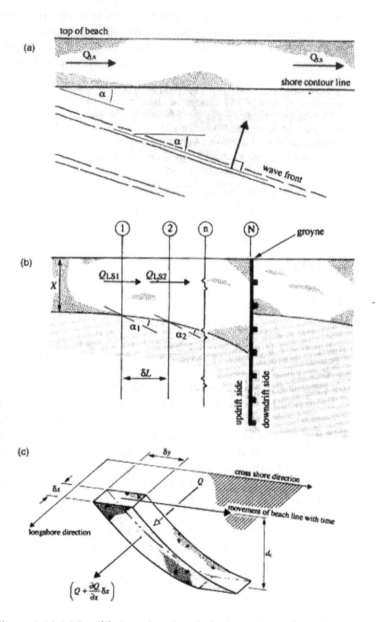

Figure 6.44 (a) Equilibrium plane bench (b) Accretion and erosion near a groyne (c) Definitions for the 1-line model.

assumed that there is no cross-shore transport into or out of the profile. Substituting in Equation (6.42), this gives

$$\frac{\partial Q}{\partial x} = d_c \frac{\partial y}{\partial t} \tag{6.43}$$

To make further progress, a relationship between the wave conditions and sediment transport is required. One well-known formula is given by the CERC equation (CERC 1984):

$$Q = Q_0 \sin 2\alpha_b \tag{6.44}$$

$$Q_0 = \varepsilon \frac{\rho}{16(\rho_s - \rho)\sigma} H_b^2 C_{gb}$$

where α is the angle between the wave crests and the shoreline, ε is a proportionality coefficient, ρ is the seawater density, ρ_s is the sediment density, σ is the sediment ratio (equal to $1 - p$, where p is the sediment porosity), H is the wave height and C_g is the wave group velocity. The angle α_b may be expressed as:

$$\alpha_b = \alpha_0 - \arctan\left(\frac{\partial y}{\partial x}\right) \tag{6.45}$$

where α_0 is the angle between the breaking wave crests and the x-axis, set parallel to the shoreline trend, and $\partial y/\partial x$ is the local shoreline orientation. The subscript b denotes these quantities at breaking.

By assuming that the wave crests approach almost parallel to the shore, the 'one-line' model reduces to a diffusion-type equation, which can be solved analytically:

$$\frac{\partial y}{\partial t} = K \frac{\partial^2 y}{\partial x^2} \tag{6.46}$$

where $K = 2Q_0/d_c$, and initial conditions and boundary conditions must be applied according to the situation being modelled (e.g., Pelnard-Considère 1956, Grijm 1961, Le Mèhautè and Soldate 1977, Larson et al. 1987, 1997, Wind 1990, Reeve 2006, Zacharioudaki & Reeve 2007, 2008) and the analytical solutions extended to the case of time-varying wave conditions and arbitrary initial beach shape.

Computational schemes solve the continuity, sediment transport and wave angle equations (Equations 6.43–6.45) simultaneously, stepping forward in time, and may include extra processes such as detailed wave transformation. They are thus better suited for engineering practice. In comparison with other means of predicting shoreline evolution, 'one-line' models are computationally cheap, have modest data requirements and have performed satisfactorily in engineering projects (e.g., Gravens et al. 1991, Reinen-Hamill 1997, Dabees and Kamphuis 1998).

6.7.2 Examples

This set of examples illustrates wider applications of a probabilistic approach to coastal management, as opposed to applications of reliability analysis. The first example concerns the effect of the construction of a groyne on a beach, and the fluctuations of the beach near the groyne in response to a varying sequence of waves. The response of the beach to a given wave condition is dependent upon the local alignment of the beach, which is a function of the earlier waves in the sequence. The beach response is thus correlated, to a degree, with the waves. Furthermore, the existence of the groyne means that the statistics of beach position are not stationary in either space or time. The second example describes the modelling of a flood defence scheme in Norfolk, UK. In this scheme, a set of shore-parallel detached breakwaters have been constructed in an attempt to hold the beach in place to prevent flooding of valuable habitat in the hinterland. Not only are the statistics of the beach nonstationary in time, but there are additional physical processes that need to be accounted for in transforming the waves from deepwater to the shoreline. This entails numerical modelling of the wave propagation and thus the use of a Level 3 Monte Carlo simulation to determine the statistics of the beach evolution near the breakwaters. The final example illustrates the application of the re-sampling technique of jack-knifing to the prediction of the movement of offshore sandbanks. Rather than using a computational dynamic model, a sequence of seabed measurements are used to define spatial and temporal patterns of change in the sandbank configuration. These patterns are extrapolated into the future, using a jack-knife method, to create an ensemble of possible futures from which the mean and variance of future seabed levels may be estimated.

Example 6.24. Description of the statistics of beach position near a groyne

Statement of problem – Groynes are often employed as part of coast protection schemes, in an attempt to hold the beach in position, thereby providing the ability to dissipate the incoming wave energy before it reaches the upper shore. Due to the variability in wave conditions, the exact height and position of the beach will vary over time, thereby providing a variable level of protection. You are provided with a time history of wave conditions and asked to investigate the behaviour of the expected position of the beach position in the vicinity of a groyne, assuming the beach is initially straight.

No *failure criterion* is defined explicitly, but one could imagine that a possible criterion might be that the beach contour of interest should remain a particular distance seaward of the upper beach. The *basic variables* are the beach position and the wave variables. We are given a large amount of wave data, but little in the way of explicit statistical information. After some preparatory work it is likely that the *method of solution* will be based on analysing the nature of the beach response to the stochastic wave conditions.

Solution. This example is taken from recent work (Pedrozo-Acuña et al. in press) on quantifying the uncertainty of beach position near groynes. The thrust of the investigation is to compare approximate analytical forms of the ensemble mean beach position against the solution obtained from Monte Carlo simulation. The ensemble mean beach position is a function of along-shore distance and time. The analytical solution described by Reeve (2006) is used as the basis of the computations. The study is based on data relating to Cardigan Bay, Wales, and the beach at Aberystwyth.

A time series of 3-hourly wave conditions in the bay in deep water are transformed to shallow water at the Aberystwyth frontage using the SWAN wave transformation model (Booij et al. 1996). Figure 6.45 shows the summary statistics of the transformed waves.

In order to estimate ensemble averaged quantities, multiple realisations of wave sequences with the correct probability distribution and correlation properties are required. The time series of waves may be considered a single realisation of wave conditions. This can be used to generate more sequences with similar statistical properties, in order to generate an ensemble of solutions. The procedure described by Scheffner and Borgman (1992) was used to generate such an ensemble. This consists of a piecewise, month-by-month, multivariate, stationary simulation approach, which preserves the marginal distributions and the first- and second-order moment properties that describe the intercorrelations within the data sequence. Seasonal changes are captured by simulating each month separately based on the information of the original time series for each month. A smooth transition in the time series and intercorrelations from month to month are enforced by interpolation.

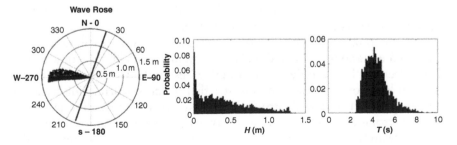

Figure 6.45 Wave rose and probability distributions of wave height and period of waves near Aberystwyth.

Each realisation within the ensemble of sequences may be used to drive a one-line model to predict the evolution of the beach of the same time period. Statistics of the beach position at any position or time can be computed in the standard manner from the sample of realisations (see Section 5.7).

For an initially plane beach with a groyne, the analytical solution for the beach contour position, $y(x, t)$, takes the following form:

$$y = -\frac{1}{\sqrt{\pi}} \int_0^t T1\,T2\,T3\ dw \tag{6.47}$$

where $T1$, $T2$ and $T3$ are (complicated) functions of x and w, and w is a dummy variable of integration (see Reeve 2006). This expression can be evaluated using numerical integration for a given sequence of wave conditions. An expression for the ensemble average beach position can be written down directly from Equation (6.47) as:

$$\langle y \rangle = -\frac{1}{\sqrt{\pi}} \int_0^t \langle T1\,T2\,T3 \rangle\ dw \tag{6.48}$$

where $\langle \bullet \rangle$ denotes the ensemble average and is the triple integral $\int_0^\infty \int_0^\infty \int_0^{2\pi} d\theta\,dT\,dH$ over all possible values. This is difficult to treat analytically, because it involves a cross-correlation of three terms. If the three terms are approximately statistically independent, then we may approximate the ensemble of the product of functions by the product of the ensemble averages,

$$\langle y \rangle \approx -\frac{1}{\sqrt{\pi}} \int_0^t \langle T1 \rangle \langle T2 \rangle \langle T3 \rangle\ dw \tag{6.49}$$

For this case, Pedrozo-Acuña et al. (2009) have demonstrated that Equation (6.49) is a very good approximation.

Figure 6.46 shows the comparison of the ensemble average solution evaluated from Equations (6.48) and (6.49) at a point 4 years into the time series. The agreement between the results obtained using the two methods is very good, at least for this case, and indicates that the cross-correlation between the three terms is not at all strong. This demonstrates that, although there may be strong correlations between, say, wave height and wave direction, this correlation appears not to be very important in the context of the morphological evolution of the beach plan shape.

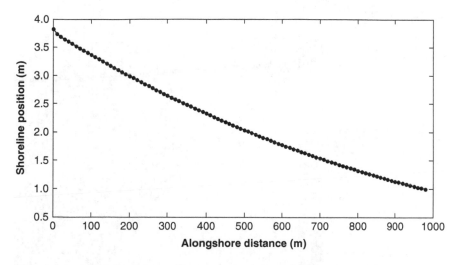

Figure 6.46 (solid line) Ensemble average of the solution estimated using Equation (6.48); (dots) ensemble average of the solution estimated using Equation (6.49).

Example 6.25. Monte Carlo simulation beach response around a detached breakwater scheme

Statement of problem – The Happisburgh to Winterton sea defences currently provide protection against sea flooding to a large part of the hinterland (including the Norfolk Broads), freshwater-dependent habitats, as well as people and properties (see Figure 6.47). Over the period of several centuries the 14 km of Norfolk coastline between Happisburgh and Winterton has suffered a series of serious floods, resulting in severe damage to property, agricultural land and the loss of many lives. Historic records show that, in 1287, nearly 200 people were drowned when extensive flooding reached as far inland as Hickling. In 1604, a flood inundated some 800 ha of land at Eccles, destroying over 60 houses and badly damaging the town church. In more recent times, the sea has breached the dunes twice at Horsey. In 1938, the whole village was cut off and an area of 3000 ha was flooded, with consequent disruption to the community and damage to agricultural land. On the night of 31 January 1953, a severe north-westerly gale, coincident with a period of spring tides, produced a great tidal surge which raised the sea level to 2.4 m above normal high-tide levels. This washed away the dunes at Sea Palling, causing extensive damage to houses and the death of seven people. The defences at Horsey and Eccles were also damaged and an area of almost 500 ha was inundated.

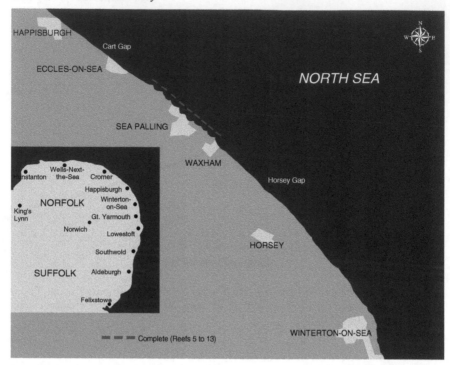

Figure 6.47 Location map of the Happisburgh to Winterton scheme.

An investigation of this disaster by the government led to the publication of the Waverley Report in 1954. This advised that the high water level of the 1953 surge should be taken as the maximum against which protection could reasonably be afforded. Subsequently, sea defences along the coast were improved so that the coastline between Happisburgh and Winterton was protected by a continuous concrete seawall, the height of which was consistent with the worst recorded tide level. The integrity of this wall depends on its foundations being protected from wave attack by beach sand, especially under storm conditions. The maintenance of adequate beach levels is therefore essential to continued sea defence.

The existing seawall and dunes protect a large area of low-lying hinterland from flooding during storm events. A significant proportion of the coastline is of national importance for its landscape and has been designated an Area of Outstanding Natural Beauty (AONB). Much of the inland area is within the Broads Environmentally Sensitive Area, and is recognised for its important landscape, historical interests and wildlife, receiving similar status to a National Park. A breach in the defences would cause extensive damage to properties, agricultural land and these sites of nature-conservation importance. Some estimates suggest that 6000 ha of fertile productive agricultural land would be turned into saltmarsh and seawater would flood into the Norfolk Broads.

By the late 1980s the beach in front of the seawalls had lowered to levels that were considered critical. To address this, the government sanctioned the construction of nine offshore reefs (Reefs 5–13). These were constructed in two stages: Reefs 5–8 in Stage 1 (1993–1995) and the remaining reefs in Stage 2 (1997) of the project, during which time beach recharge was undertaken. The reefs constructed in Stage 2 used a modified version of the Stage 1 reef design, which had lower crest levels, were shorter in length and had more closely spaced reefs. In addition to Stages 1 and 2 construction, some intermediate works have been undertaken to ensure the integrity of existing defences where beach loss has occurred in areas not defended by the reefs. From 2002 to 2004, Stage 3 works were completed and comprised beach recharge, the construction of a rock revetment south of Reef 13 and the improvement of rock groynes between Cart Gap and Bramble Hill. Reefs 1–4 were reserved for a subsequent stage and have yet to be constructed; thus, the northernmost reef is Reef 5.

In this case we have a complicated system but the *failure criterion* could be defined either as the inundation of all or part of the area currently protected by the coastal defences, or as the undermining and collapse of the flood defence structures at the top of the beach which, if left unrepaired, would lead to inundation. In either case, a key part of the problem is to determine the likely excursions of the beach under varying wave and tide conditions. The *basic variables* are the beach position, water levels and wave variables. We are given a long sequence of deep-water wave conditions and water level recordings. Waves propagating towards the shore will undergo shoaling, refraction breaking and diffraction (around the ends of the breakwaters). Furthermore, waves will break on and/or over the breakwaters. The amount of wave overtopping of the breakwaters will affect sediment transport and thus the movement of the beach. This situation is too complicated for treatment with an existing analytical solution, so Monte Carlo simulation (a Level 3 method) is a possible way forward. After some preparatory work, the *method of solution* will be based on generating realisations of beach response and analysing these to determine the appropriate statistics. An oblique aerial photograph of the scheme taken towards the end of the construction of Stage 2 is shown in Figure 6.48, viewing the scheme from the northern end looking southwards.

Solution. This example is taken from research undertaken as part of a project funded by the UK Research Council, a collaborative venture between the Universities of Liverpool, East Anglia and Plymouth, the Proudman Oceanographic Laboratory and the British Oceanographic Data Centre (BODC). Industry representatives Halcrow Group and HR Wallingford are also partners in the project, which is funded in parallel by the UK Department for Environment Food and Rural Affairs (DEFRA) and the

Figure 6.48 Oblique aerial photo of the Happisburgh to Winterton scheme, viewing
    south-eastwards (with courtesy of Mike Page).

UK Environment Agency (EA) to incorporate project results into national
design guidelines. The project includes an extensive campaign of field
measurements, detailed process modelling and probabilistic shoreline modelling for medium- to long-term beach evolution prediction. Further details
can be found at http://pcwww.liv.ac.uk/civilCRG/leacoast2/index.htm.

To drive the beach evolution model, nearshore waves are required. The
offshore wave data are the output from the UK Meteorological Office's
European wave model over the period 1995–2008. The mean water depth
is 18 m and the wave height, period and direction are available at 3-hourly
intervals. Figure 6.49 shows the wave rose of this data, while Figure 6.50
shows a segment of the time series of wave heights, periods and directions,
in which an annual signal is evident in the wave heights. The deep-water
waves are transformed using a numerical wave model that accounts for
refraction, diffraction and breaking.

The 13-year wave record was used to create 200 more 13-year sequences
with similar statistical properties, using the method described by Cai et al.
(2008). In Monte Carlo simulation, creating realisations with the correct
statistical properties is a crucial part of the process, and the realisations
must be recreated accurately. In particular, this is true for the marginal

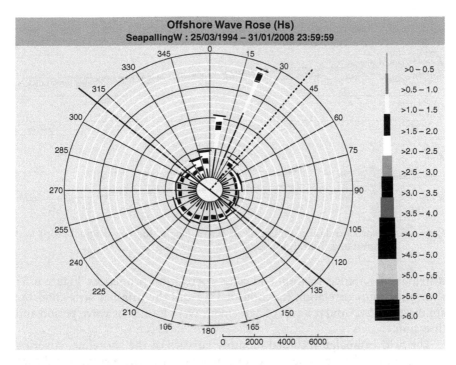

Offshore Wave Rose (Hs)
SeapallingW : 25/03/1994 – 31/01/2008 23:59:59

Figure 6.49 Offshore wave rose of hindcast data. The full line shows the trend of the nearby coastline.

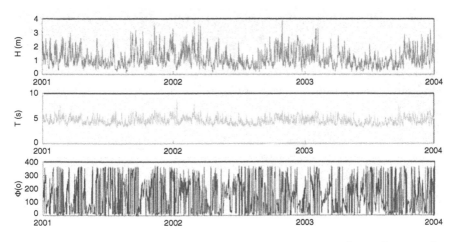

Figure 6.50 Segment of the time series of deep-water wave conditions between the beginning of 2001 and the end of 2003.

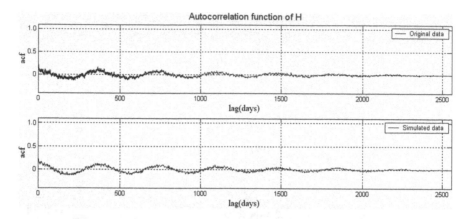

Figure 6.51 Autocorrelation functions of wave height for the original (top) and simulated (bottom) series.

distributions, autocorrelation and cross-correlation properties. Figure 6.51 shows the autocorrelation function of H_s from the original series and the simulated ones. Similar levels of agreement were found for wave period and direction.

The field campaign included monthly surveys of the shoreline, covering the area of the defence scheme. An example of one of these surveys is shown in Figure 6.52.

The first of these was used to define an initial shoreline condition for a one-line model. A wave model based on the mild-slope equation model

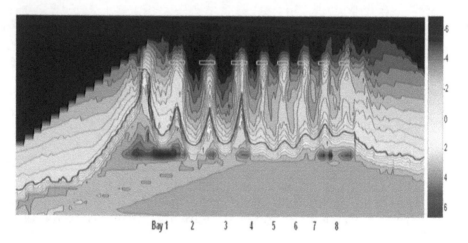

Figure 6.52 Coastal bathymetric and topographic survey. The heavy contour shows the mean water line and numbering scheme of the individual bays formed by the breakwaters.

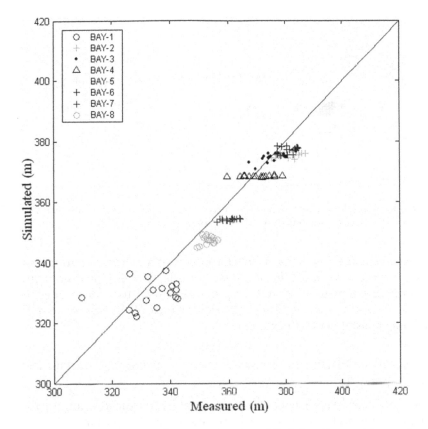

Figure 6.53 Comparison of measured and modelled shoreline position at the centre of the bays.

described by Li (1994) was used to transform the sequences of waves from deep water towards the shore on a regular grid with resolution 10 m. The transformed waves were then used to drive the one-line model. The positions of the centres of each of the bays corresponding to the survey dates were output from the model. Figure 6.53 shows the comparison between the calibrated model predictions and the measured mean water line.

Most of the predictions are within about 5 m of the measured position, with one or two outliers, demonstrating a good level of agreement for this type of model. The calibrated model was then run repeatedly with the 200 other 13-year realisations to create an ensemble of possible outcomes. Some summary statistics from this output are shown in Figure 6.54.

This indicates that the shoreline behind the Stage 1 breakwaters is relatively stable, with the exception perhaps of the Bay 1. From Reef 8 and throughout Stage 2 the shoreline shows a greater variability in position of between 20 m and 50 m in the horizontal.

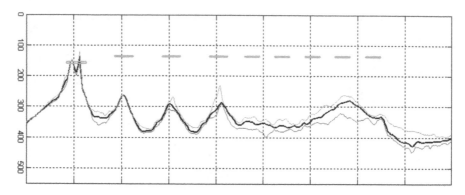

Figure 6.54 Ensemble average shoreline (heavy line) with extreme positive and negative excursions over a 13-year period.

These particular simulations include the effects of tide-level variation, but not longer-term water-level changes due to sea-level rise. However, maps such as the one shown in Figure 6.54 provide a useful tool for local engineers and planners when assessing the performance of the scheme and in formulating coastal management strategy.

Example 6.26. Offshore sandbanks at Great Yarmouth – a statistical ensemble approach

Statement of problem – The offshore sandbanks at Great Yarmouth have major environmental, commercial, leisure and physical importance. An offshore wind farm is located on top of one of them (Scroby Sands – see Figure 6.55); the access to the newly extended port at Great Yarmouth relies on clear channels running through them; they are used as breeding grounds for North Sea seals; and they provide protection to the coast from wave action.

Local knowledge has suggested that significant changes to the bank and channel configuration can occur over a period of several years. The main banks in the system are shown in Figure 6.56.

Any significant alterations in the configuration of the sandbanks would undoubtedly have an effect on the nearby shoreline and associated shoreline management. The ability to forecast any such changes is of direct relevance to the strategic management of this shoreline. The difficulty in doing so is that: (a) the computing power required to simulate the tides, waves and sediment transport over such a large area for the order of several years is simply prohibitive; and (b) even if one could do the simulations, the uncertainties associated with initial conditions mean that long-term forecasts with detailed dynamical models are very unreliable. As a long sequence of 36 bathymetric charts from 1848 to 2008 is available, an alternative is to use

Figure 6.55 Oblique aerial photograph (viewing northwards) of Scroby Sands in 2006 (courtesy of Mike Page).

a data-driven modelling approach (e.g., Habib and Mesellie 2000, Różyński et al. 2001, Różyński and Jansen 2002, Różyński 2003). These methods rely on some form of statistical analysis of observations and extrapolation into the future, but suffer from the constraint that if a particular behaviour is not contained in the observations it is unlikely that the extrapolation technique will accurately predict such a change. It does, however, have the advantage that it is ideally suited to quantifying the uncertainty of forecasts (Larson et al. 2003).

This complicated morphological system could be considered to fail in terms of a structure if its configuration evolved in such a way that the wind farm was exposed to larger waves; if the channels between the banks infilled sufficiently to cause navigation difficulties for ships using the port; or if they eroded to an extent that seals could no longer use them as a breeding ground; and also that greater wave energy could reach the shoreline. A *failure criterion* could be defined in terms of any of these conditions. The *basic variables* are not only the total volume of material in the sandbanks but also its spatial arrangement. Indirectly, the tides and waves are also variables, as these determine sediment transport, and ultimately the changes in sandbank configuration. We are given a long sequence of bathymetric surveys. This situation is too complicated for treatment with existing process models, and

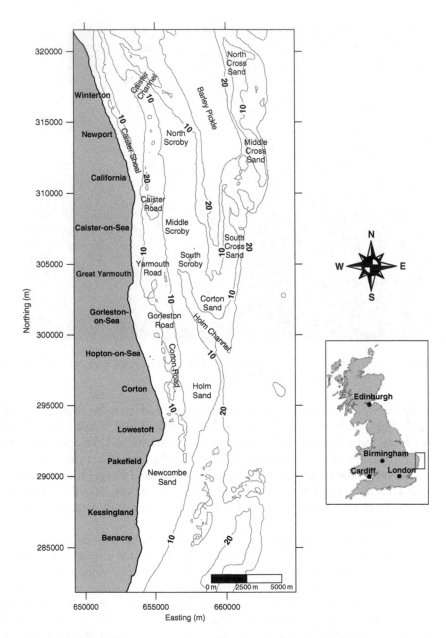

Figure 6.56 The Great Yarmouth sandbank system and coastline.

data-driven modelling offers an alternative. However, a measure of uncertainty in the prediction is required, so some means of estimating the error or uncertainty in the predictions is required. A re-sampling (Monte Carlo type) approach could provide this. The *method of solution* is based on a purely statistical approach, with no physical process modelling.

Solution. This example follows the paper by Reeve et al. (2008) and details of the approach may be found in that paper and references therein. Here, a brief outline is given, together with some of the results.

The steps in the analysis are:

a   The bathymetric measurements up to and including 1998 were used to perform a set of empirical orthogonal function (EOF) analyses. EOFs were introduced to coastal engineering by Winant et al. (1975) who used them to analyse changes in beach profiles. The EOFs provide the means of expressing the patterns of spatial and temporal change in the data as a series of functions. It is a very efficient way of compressing variability in the data into a small number of functions (usually much more efficient than performing a Fourier expansion). Each EOF describes a spatial distribution of variation. For each EOF there is a corresponding function that describes its variation in time. The whole data set of bathymetric charts can be expressed as a sum of EOFs.

b   There are 33 charts for the period 1848–1998. Rather than performing a single analysis of all the charts, the sequence of charts is used to create an ensemble of EOF analyses by using the jack-knife technique: 33 EOF analyses are performed, leaving out each one of the charts in turn. This introduces an uncertainty into the EOFs equivalent to that associated with the loss of one of the charts.

c   The corresponding temporal functions obtained from these analyses were extrapolated to 2006 using a technique that mimics the past temporal behaviour (an autoregressive prediction procedure).

d   The extrapolated temporal functions were then recombined with their corresponding spatial functions to create an ensemble of forecasts of the bathymetry in 2006, that is, a prediction 8 years into the future from the date of the last chart used in the EOF analysis.

e   These forecasts were then compared to the chart for 2006, and ensemble statistics computed as in Section 3.7.

Some of the results are shown in Figure 6.57(a–e). The general trends are that the average of the ensemble agrees fairly well with the observed. Agreement is less good in the more seaward banks and Holm Channel. One drawback of any data-driven method is that there is no explicit inclusion of physical processes in the forecasting, so it cannot capture

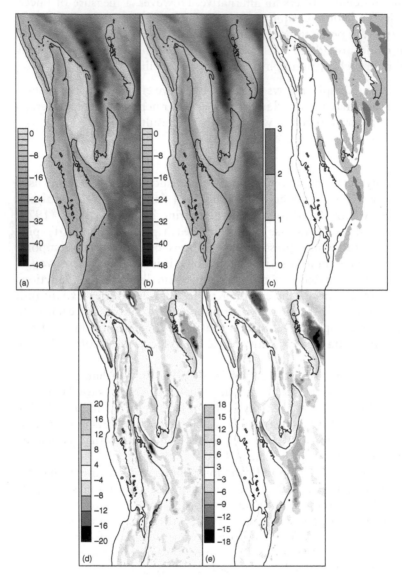

Figure 6.57 Forecasts to year 2006: (a) bathymetry of 2006; (b) mean bathymetry of the ensemble forecasts; (c) standard deviation of the ensemble of forecasts; (d) difference of 2006 bathymetry and ensemble mean forecast bathymetries in units of standard deviation; (e) difference of 2006 bathymetry and the ensemble mean forecast. The 5 m depth contour for the measured 2006 bathymetry is superimposed on plots (b–d) for reference (Reprinted *from Estuarine Coastal and Shelf Science*, Vol. 79(3), Reeve, Horrillo-Caraballo & Magar, Statistical analysis and forecasts of long-term sandbank evolution at Great Yarmouth, UK, Copyright [2008], with permission from Elsevier).

changes due to processes that are not reflected in the historical record. However, the method provides forecasts and uncertainty bounds, but has some deficiencies associated with the choice of expansion functions (EOFs), and could probably be improved in terms of the form of extrapolation used.

Further reading

Ang, A. H.-S. and Tang, W. H., 1975. *Probability Concepts in Engineering Planning and Design, Volume I: Basic Principles,* John Wiley and Sons, New York, NY.

Christensen, B. A., 2000. Discussion on "Limitations and proper use of the Hazen-Williams equation," by C. P. Liou, *Journal of Hydraulic Engineering, ASCE,* 126(2): 167–168.

Churchill, S. W., 1973. Empirical expressions for the shear stress in turbulent flow in commercial pipe. *Journal of AIChE,* 19(2): 375–376.

Cowell, P. J., Stive, M. J. F., Niederoda, A. W., De Vriend, H. J., Swift, D. J. P., Kaminsky, G. M. and Capobianco, M., 2003a. The coastal-tract (part 1): a conceptual approach to aggregated modelling of low-order coastal change. *Journal of Coastal Research,* 19(4): 812–827.

Cowell, P. J., Stive, M. J. F., Niederoda, A. W., Swift, D. J. P., De Vriend, H. J., Buijsman, M. C., Nicholls, R. J., Roy, P. S., Kaminsky, G. M., Cleveringa, J., Reed, C. W. and De Boer, P. L., 2003b. The coastal-tract (part 2): Applications of aggregated modelling of lower-order coastal change. *Journal of Coastal Research,* 19(4): 828–848.

Cowell, P. J., Thom, G., Jones, A., Everts, C. H. and Simanovic, D., 2006. Management of uncertainty in predicting climate-change. *Journal of Coastal Research,* 22(1), 232–245.

Davison, A. C. and Hinkley, D. V., 1997. *Bootstrap Methods and Their Application,* Cambridge University Press, Cambridge, p. 592.

EurOtop Manual, 2008. Downloadable from http://www.overtopping-manual. com/, accessed 06/02/09.

Hotelling, H., 1933. Analysis of a complex of statistical variables into principle components. *Journal of Educational Psychology,* 24: 417–441.

Howd, P. A. and Birkemeier, W. A., 1987. *Beach and Nearshore Survey Data: 1981–1984.* CERC, Field Research Facility. Tec. Rep. CERC-87-9, U. S. Army Eng. Waterways Expt. Stn., Coastal Eng. Res. Ctr., Vicksburg, MS.

Jenkinson, A. F., 1955. The frequency distribution of the annual maximum (or minimum) value of meteorological elements. *Quarterly Journal of the Royal Meteorological Society,* 81: 158–171.

Li, Y., Lark, M. and Reeve, D. E., 2005. The multi-scale variability of beach profiles at Duck, N. C., USA, 1981–2003. *Coastal Engineering,* 52(12): 1133–1153.

Liou, C. P. 1998. Limitations and proper use of the Hazen-Williams equation. *Journal of Hydraulic Engineering, ASCE,* 124(9): 951–954.

Marple, S. L., Jr., 1987. *Digital Spectral Analysis with Applications,* Prentice Hall, Englewood Cliffs, New Jersey, USA.

Merrifield, M. A. and Guza, R. T., 1990. Detecting propagating signals with complex empirical orthogonal functions: A cautionary note. *Journal of Physical Oceanography,* 20: 1628–1633.

Miller, J. K. and Dean, R. G., 2007. Shoreline variability via empirical orthogonal function analysis: Part I temporal and spatial characteristics. *Coastal Engineering*, 54(2), 111–131.

Naess, A., Gaidai, O. and Haver, S., 2007. Efficient estimation of extreme response of drag-dominated offshore structures by Monte Carlo simulation, *Ocean Engineering*, 34: 2188–2197.

North, G. R., Bell, T. L. and Cahalan, R. F., 1982. Sampling errors in the estimation of empirical orthogonal functions. *Monthly Weather Review*, 110: 699–706.

Percival, D. B. and Walden, A. T., 1993. *Spectral Analysis for Physical Applications: Multitaper and Conventional Univariate Techniques*, Cambridge University Press, Cambridge, UK.

Pourahmadi, M., 2001. *Foundations of Time Series Analysis and Prediction Theory*. John Wiley & Sons, New York, p. 448.

Reeve, D. E. and Spivack, M., 2004. Probabilistic methods for morphological prediction. *Coastal Engineering (Special Issue on Coastal Morphodynamic Modelling)*, 51(8–9): 661–673.

Reeve, D. E., Li, B. and Thurston, N., 2001. Eigenfunction analysis of decadal fluctuations in sand bank morphology at Gt Yarmouth. *Journal of Coastal Research*, 17(2): 371–382.

Robinson, A. H. W., 1960. Ebb-flood channel systems in sandy bays and estuaries. *Geography*, 45: 183–199.

Robinson, A. H. W., 1966. Residual currents in relation to shoreline evolution of the East Anglian Coast. *Marine Geology*, 4: 57–84.

Schoones, J. S. and Theron, A. K., 1993. Review of the field-data base for longshore sediment transport. *Coastal Engineering*, 19: 1–25.

Silvester, R. and Hsu, J. R. C., 1997. *Coastal Stabilisation*, Advanced Series on Ocean Engineering, Vol. 14, World Scientific, Singapore.

Tung, Y. K. and Yen, B. C., 2005. *Hydrosystems Engineering Uncertainty Analysis*, McGraw-Hill Book Company, New York, NY.

Van der Burgh, L. M., Wijnberg, K. M., Hulscher, S. J. M. H., Mulder, J. P. M. and van Koningsveld, M., 2007. Linking Coastal Evolution and super storm dune erosion forecasts. In Kraus, N. C. and Rosati, J. D. (Eds) Proceedings of the 6th International Symposium on Coastal Engineering and Science of Coastal Sediment Processes, New Orleans, Louisiana, May 13–17, 2007. ASCE, 0784409269, 2007, p. 2640.

Van der Meer, J.W., 1988a. Deterministic and probabilistic design of breakwater armour layers. *Proc. ASCE, Journal of Waterway, Port, Coastal and Ocean Engineering*, 114(1): 66–80.

Van der Meer, J.W., 1988b. *Stability of Cubes, Tetrapods and Accropodes, Design of Breakwaters*, Thomas Telford, London, UK.

Van der Meer, J. W., 1988c. *Rock Slopes and Gravel Beaches Under Wave Attack*, PhD Thesis, Delft Technical University, April 1998. (Published as Delft Hydraulics Commn. No. 396).

Van der Meer, J. W., 1990. Static and dynamic stability of loose materials, in *Coastal Protection* (Ed. K. W. Pilarczyk), Balkema, Rotterdam.

Van der Meer, J. W. and Koster, M. J., 1988. Application of computational model on dynamic stability. Proceedings Conference on Breakwaters – Design of Breakwaters, Thomas Telford, London.

Van Goor, M. A., Stive, M. J. F., Wang, Z. B. and Zitman, T. J., 2001. Influence of relative sea level rise on coastal inlets and tidal basins. Proceedings Coastal Dynamics 2001, ASCE, Lund, Sweden, pp. 242–251.

Weibull, W. 1939. A statistical theory of the strength of materials. Proceedings Royal Society Swedish Institute for Engineering Research, No 151.

Zyserman, J. A. and Johnson, H. K., 2002. Modelling morphological processes in the vicinity of shore-parallel breakwaters. *Coastal Engineering*, 45: 261–284.

7   Reliability of systems

7.1   Introduction

So far only individual elements and individual modes of failure have been
considered. In practice, a coastal defence or hydraulic structure will have
more than one failure mechanism and will often be part of a larger scheme
that may contain several or many linked components. The entirety of a
drain is a good example, as it may contain not only pipe sections but
also joints, junctions, inflows, outflows and stilling tanks, to name but a
few components. In a similar vein, a flood defence is likely to comprise
several different types of construction along its length, as well as flood
gates and barriers. In some cases where there is a change in construction the neighbouring structure provides a key element of stability, so that
should it fail then its neighbour would be much more likely to fail too.
Such an example would be vertical steel sheet piles, topped with a concrete beam or pad, located next to a mass concrete structure. In the event
that the mass concrete structure failed (most likely through gradual deterioration and lack of maintenance), the sheet piling would be at greater
risk because water could attack from both front and rear, washing out
fines from behind the piles and reducing the integrity of the structure as a
whole.

To simplify the study of system reliability two categories of systems are
considered. In a *series system*, the failure of one member automatically leads
to the failure of the whole system. An example of such a system is a chain,
which fails when the weakest link breaks. In a *parallel system*, all the members must fail before the system fails. An example of this type of system is a
cable composed of several strands of wire. The cable does not fail until all
the component strands have failed.

When considering a hydraulic structure it can be helpful to consider it
as a system comprising a number of elements. Certain elements may fail
without causing the system as a whole to fail, whereas if other elements fail
the whole system fails. In the former case the elements may be considered
to be connected in parallel in analogy to connections in an electrical circuit.
In the latter case the elements are connected in series.

Table 7.1 Simple bounds for systems

	Lower bound	Upper bound
Series system	$\max\limits_{i=1,n} P_i$	$1 - \prod\limits_{i=1}^{n}(1 - P_i)$
Parallel system	$\prod\limits_{i=1}^{n} P_i$	$\min\limits_{i=1,n} P_i$

Now, each element will have a certain probability of failing. Let the probability of failure for element i be P_i. Upper and lower bounds on the probability of system failure can be obtained by assuming that (a) the elements are all perfectly correlated and (b) there is no correlation between any pair of elements; the bounds and are summarised in Table 7.1.

For series systems the lower bound corresponds to the assumption of perfect dependence (i.e., if one element fails they all do), while the upper bound corresponds to the assumption of no pairwise correlation between elements (see, for example, Equation 4.1). For parallel systems the lower bound arises from the assumption of independence and is the probability of the occurrence of n independent events. The upper bound arises from the assumption of perfect dependence. Note that in this treatment of parallel systems issues regarding the (time-dependent) spreading of the additional load amongst the remaining functioning elements is not considered. In practice these bounds may be so wide as to provide no useful constraint. Other bounds have been suggested in the literature, such as those proposed by Ditlevsen (1979) for series systems.

7.2 Overview of methods

7.2.1 Single structure, multiple mutually independent failure mechanisms

For a series system a *fault tree* may be used to analyse the reliability of each element and unit of the system in a logical manner. In its strictest definition, a fault tree is a description of the logical interconnection between various component failures and events within a system. A fault tree is constructed from events and gates. Gates are logical operators used to combine events to give an event at a higher level. Gates are built from the logical operators (sometimes known as Boolean operators), AND, OR and NOT. The highest level of event is known as the TOP event, which would normally be chosen to correspond to failure of the system as a whole.

The main purpose of constructing a fault tree is to establish the logical connection between different sets of component failures, assign values to component reliability and thus calculate the probability of the TOP event occurring.

Fault tree analysis is strictly only applicable to systems in which components can be in one of two states: working or failed. Where the processes studied do not obey this behaviour, fault trees on their own are less appropriate. Coastal defences are a good example of such systems. Nevertheless, fault tree analysis can be a useful step in a qualitative assessment of risk. Figure 7.1 shows the fault tree concept applied to a flood defence system that contains four elements. Starting with the event of system failure and working backwards, the possible precursors of this event are identified. The tree can be extended further downward from system level to element level to unit level and so on.

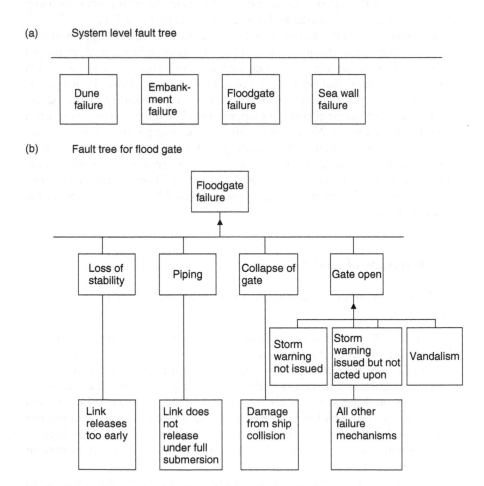

Figure 7.1 (a) System level fault tree. (b) Fault tree for flood gate.

The power of the fault tree is evident in Figure 7.1, where a floodgate is analysed. A number of technical failure mechanisms may be identified for which design equations are available. This, together with knowledge of the environmental and design conditions, enables the probability of these events to be estimated using the techniques described in earlier chapters. In the case of a floodgate a non-negligible contribution to the probability of failure arises from human error or intervention. The reliability of this element of the system is influenced by the possibility of management failure, and this cannot be expressed by classical engineering calculations. The fault tree at least identifies this issue and illustrates how it fits into the reliability of the system as a whole. Fault trees for excessive waves behind a rubble breakwater and inundation due to failure of a flood defence have been developed by CIAD (1985) and CIRIA & CUR (1991), respectively. The calculation of probabilities of failure from fault trees usually requires some assumptions about the mutual exclusivity of failure modes (i.e., failure arises from one or other failure mechanism, not a combination of them) and independence between members of the system.

Example 7.1. *Statement of problem* – A spherical buoy (with instrumentation) is connected to the seabed with a rope and tension-release link. The link is designed to release when the buoy is semi-submerged. Four failure mechanisms have been identified and are shown in the corresponding simple fault tree.

These are considered to be mutually exclusive failure mechanisms. They are also considered exhaustive: by virtue of the fact that all other possible failure mechanisms not identified explicitly are included in the category 'all other failure mechanisms'. As a result, once the probability of occurrence of each failure mechanism has been determined the total probability of failure will be given by the sum of the probabilities of each mode of failure. Furthermore, the probability of non-failure is (1 – the total probability of failure). For the present we are given that the last two probabilities are 0.001 and 0.002, respectively. The performance of the particular tension-release mechanism is uncertain, but tests on multiple batches of the mechanism have shown that the actual release tension, T_r, is known to follow a uniform distribution between $0.7\,T_c$ and $2.1\,T_c$, where T_c is the target release tension $= mg - \rho_s(2/3)\pi r^3$ where m and r are the mass and radius of the buoy, and ρ_s is the density of seawater. Determine the probability that the buoy system will fail.

The *failure criterion* is fourfold. First, the system may fail if the buoy is released before becoming semi-submerged ($T_r < T_c$); second, the buoy may fail to release even under full submersion ($T_r > 2T_c$); third, the buoy may be damaged by ship collision; and, finally, the system may fail through some other unspecified mechanism.

The *basic variable* is the release tension, T_r.

This is a problem concerning Level 2 methods, as we are given a probability distribution. However, due to the simple nature of the system a solution can be found directly.

Solution. Let P_{fi} denote the probability of failure due to the ith failure mode. Then $P_{f3} = 0.001$ and $P_{f4} = 0.002$. Now,

$$P_{f1} = P(T_r < T_c) = 3/10 \times 5/7 = 3/14, \text{ and}$$
$$P_{f2} = P(T_r > 2T_c) = 1/10 \times 5/7 = 1/14$$

As the modes of failure are mutually exclusive, the total probability of failure is given by the sum of the probabilities of each mode of failure, that is, the total probability of failure is $3/14 + 1/14 + 0.001 + 0.002 = 0.289$.

Example 7.2. *Statement of problem* – Consider an embankment that provides defence against flood-water levels corresponding to the 1 in T year event. Suppose also that the structure has a design life of N years, during which it is maintained to the standard of its original construction. What is the probability that the embankment will be overcome by an extreme flood event? The *failure criterion* is simply that the water level, W, exceeds the 1 in T year level, W_T. The *basic variable* is water level, but the nature of the problem is closer to the ideas covered in Chapter 2 than with the reliability theory covered in Chapter 5.

Solution. The question is equivalent to the question, 'Will the 1 in T year water level be exceeded in the next N years?' To see how the solution can be derived we begin by looking at the cases $N = 1$ and $N = 2$, and then generalise.

For $N = 1$: the probability that the water level exceeds the 1 in T year level is $p \equiv P(W > W_T) = 1/T$, and the probability that it does not is $P' = 1 - p$.

For $N = 2$: we treat each year as an independent event. Thus, the probability that the water level does not exceed the 1 in T year level in either the first or second year is the product of the probabilities that it does not exceed it in the first year and the second year, that is, $P(W \leq W_T \text{ in year 1 and } 2) = (1 - p)(1 - p)$.

Generalising, for N years the probability that the water level does not exceed the 1 in T year level is $(1 - p)^N$. Hence, from the theorem of total probability (Chapter 2), the probability that the water level does exceed W_T during N years is $1 - (1 - p)^N$.

As an illustration, Table 7.2 shows the probability of failure for different design lives with $T = 100$ years.

Table 7.2 Failure probabilities for Example 7.2

Design life (years)	Probability of failure (%)
1	1
5	5
50	39.5
100	63.4
500	99.3

The result can be generalised to $m = 1, 2, \ldots, M$ modes of failure. Under the assumption that the modes of failure are mutually exclusive in a series system, the overall probability of failure over the course of N years, $P_{f,N}$, is:

$$P_{f,N} = 1 - \sum_{m=1}^{m=M} (1 - p_{m,N}) \tag{7.1}$$

If all the probabilities of failure $p_{m,N}$ are identical ($= 1/T$), then Equation (7.1) may be written as:

$$P_{f,N} = 1 - M\left(1 - \frac{1}{T}\right) \tag{7.2}$$

7.2.2 Multiple independent structures, single-failure mechanism

The situation where there are multiple but independent structures in a series system that has a single failure mode is the archetypal 'weakest link' system.

Example 7.3. *Statement of problem* – Consider a system containing n elements, each with probability of failure $= 0.05$. Determine the probability of failure for the system for $n = 1, 2, 3, 10$ and 20.

Solution. Using Table 7.1,

$$P_f = 1 - \prod_{i=1}^{n} (1 - P_{fi})$$
$$= 1 - (1 - 0.05)^n$$
$$= 1 - 0.95^n$$

The probability is derived in the following manner. Let R be the 'strength' of the entire system, and let S be the deterministic load. When the load S is applied to the system, the load S_i in each constituent element may vary from element to element. Similarly, the strength of each element, R_i, may

Table 7.3 Probabilities of failure for
Example 7.3

N	P_f
1	0.05
2	0.0975
3	0.1426
10	0.4013
20	0.642

be different. The probability of failure of the system is thus determined as follows, $P_f = P(R \leq S) = 1 - P(R > S) = 1 - P(R_1 > S_1)P(R_2 > S_2)\ldots P(R_n > S_n) = 1 - [1 - P(R_1 \leq S_1)P(R_2 \leq S_2)\ldots P(R_n \leq S_n)]$.

The required probabilities are summarised in Table 7.3.

Note that, unsurprisingly, the probability of failure increases as the number of elements increases.

7.2.3   Single structure, multiple-failure mechanisms – general

A number of difficulties arise when trying to apply standard fault analysis techniques to hydraulic structures and sea defences. In particular:

- There are generally insufficient data to assign failure probabilities to individual components with any degree of confidence, because the number of failures is very small. (Contrast this with the manufacture of printed circuit boards, for example.)
- Fault trees are essentially binary in character, that is, components either work or fail. However, components of a hydraulic structure undergo various degrees of damage in response to loading of different severity and, even when damaged, can still provide a reduced level of protection.
- The combination of probabilities of different events to obtain the probability of the TOP event requires an assumption that the events are mutually exclusive, that is, failure will occur due to only one of the branches on the tree. For example, it would be easy to assume that the failures of each element of a flood-defence system are exclusive events, which would allow the combined probability to be calculated by adding the individual probabilities. However, it is possible for the elements to fail together. For example, a river embankment and flood gate could fail simultaneously if water levels in the river rose above their crest level. More problematically, during the course of a storm, failure of individual components is often influenced by the occurrence of the failure of a different component. For example, scour at the front toe of the defence is likely to affect the likelihood of failure of the front armour layer. The

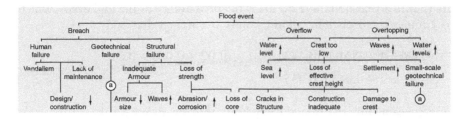

Figure 7.2 Example of a cause–consequence tree. Up and down arrows indicate conditions that are respectively larger or lower than anticipated. The 'a' items denote a link that has not been drawn with a line (adapted from Townend 1994).

fault-tree approach has difficulty in representing this behaviour. This has prompted the use of cause–consequence diagrams (e.g., Townend 1994). These show possible changes in the system and link these to consequences, thereby providing a more complete description of the system. They allow representation of some degree of recursion and dependence between chains of events. Figure 7.2 shows an example cause–consequence diagram for coastal flooding.

The problem and partial solution in this type of case is illustrated through a sequence of examples based on the failure modes of a rubble-mound breakwater (see Figure 6.15). Here, we consider only three modes of failure: m_1, excessive overtopping; m_2, armour damage; and m_3, toe scour. Furthermore, other modes of failure are excluded. In addition, M_1 denotes the event that failure mode m_1 occurs, and so on. First, we note that the sample space of failure modes $F = \{m_1, m_2, m_3, \varphi\}$, where φ is the null event (i.e., no failure). Next, recognise that *any* event affecting the structure can be written as a combination of the M_i. For example, M_2, M_1 and M_3, and $M_{1'}$ are cases where there is failure due to armour damage, failure due to overtopping and toe scour, and no failure due to armour damage. The event $M_1 \cup M_2 \cup M_3$ is the occurrence of overtopping, armour damage or toe scour, and the null event is $\varphi = \{M_1 \cup M_2 \cup M_3\}'$. To make further progress, it is necessary to assign probabilities of failure to each mode. There is an implicit time scale in this, as an 'event' has a duration. Here, the probabilities of failure are interpreted as annual probabilities. Given $P(M_1) = 0.02$, $P(M_2) = 0.005$ and $P(M_3) = 0.01$, differing relationships between the failure modes are now illustrated.

Mutually exclusive failure modes

This has been covered in Section 7.2.1 but is revisited here for the sake of completeness. In this case, the failure modes are mutually exclusive, so they cannot occur simultaneously and their intersections are null. The probability that in a year no failure occurs is the probability of the null event, which is 1 minus the probability of any failure occurring, that is,

$$P(\varphi) = 1 - P(M_1 \cup M_2 \cup M_3) = 1 - (0.02 + 0.005 + 0.01)$$
$$= 1 - 0.065 = 0.935$$

and consequently the annual probability that the defence fails is

$$P_f = 1 - P(\varphi) = 0.065$$

In this simple case, where the events are mutually exclusive the probability of any failure occurring could have been computed directly from the probability of the union of M_1, M_2 and M_3. Figure 7.3 shows a Venn diagram for this situation. Note that, because the three failure modes are mutually exclusive, there are no overlaps between any of the sets.

The probability that the structure fails during the course of 50 years due to any of the three failure modes is:

$$P_{f,50} = 1 - (1 - P_f)^{50} = 0.965$$

Note that there is no requirement for the modes of failure to be independent.

Overtopping and toe scour are not mutually exclusive

It has been observed that failure of vertical breakwaters due to overtopping occurs once scour has taken place. In this case, the conditional probability of overtopping after the toe has been scoured is not null. If $P(M_1|M_3) = 0.4$,

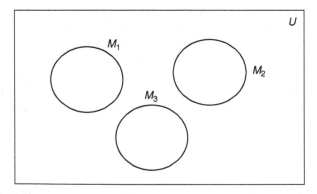

Figure 7.3 Venn diagram for mutually exclusive failure modes.

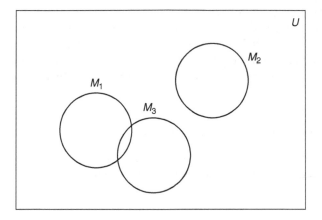

Figure 7.4 Venn diagram for the case when overtopping and toe scour are not mutually exclusive.

and the individual probabilities of overtopping, armour damage and toe scour remain as before, the probability of failure due to overtopping and toe scour is given from the conditional probability as:

$$P(M_1 M_3) = P(M_1|M_3)P(M_3) = 0.4 \times 0.01 = 0.004$$

Figure 7.4 shows the Venn diagram for this case. Note there is now an overlap between the sets of failure due to overtopping and scour, but not armour failure.

Furthermore, the probability that the defence will fail in a year due to one of the three failure modes is:

$$P_f = 1 - P(M_1 \cup M_2 \cup M_3) = \sum_{i=1}^{3} P(M_i) - P(M_1 M_3)$$

$$= 0.065 - 0.004 = 0.061$$

and the probability that the defence does not fail is $1 - 0.061 = 0.939$. The probability that the structure fails during the course of 50 years due to any of the three failure modes is:

$$P_{f,50} = 1 - (1 - P_f)^{50} = 1 - (0.939)^{50} = 0.957$$

Note that there is no requirement for the modes of failure to be independent. If, in addition, *statistical independence of overtopping and toe scour is assumed*, the previous calculations remain the same with the exception of the conditional probability. In the case when M_1 and M_3 are independent then $P(M_1|M_3)=P(M_1)=0.02$, and $P(M_1M_3)=P(M_1)P(M_3)=0.002$. Therefore, the probability that the defence will fail due to any of the three modes becomes $0.065-0.002=0.063$; the probability that it does not fail in any year is $1-0.063=0.937$ and the probability that it will fail due to one of the three modes over the course of 50 years is $1-0.937^{50}=0.961$.

None of the three failure modes are mutually exclusive

Of the cases considered this is perhaps the closest to reality. Figures 7.5 and 7.6 show damage to the seawall between the towns of Dawlish and

Figure 7.5 Erosion of material between the rail track and wave wall as a result of severe storm action (courtesy of Network Rail).

Figure 7.6 View of the damaged wall from the seaward side, showing removal of stone blocks and pointwork (courtesy of Network Rail).

Teignmouth on the south coast of the UK. At this section, the national railway line runs close to the sea, at the foot of sandstone cliffs. The railway track is protected by an almost vertical wall. There is a pedestrian walkway on the top of the wall, approximately 2 m wide, and on the landward edge there is a wave wall, approximately 1 m high. During the severe storms of October 2004, waves approached from the south/southeast rather than the more typical southwest direction. At the peak of the storm, visual estimates of the near-shore significant wave height were of the order of 3.5 m. There was a large amount of wave overtopping, which led to flooding of the railway line, closure of services for a few days and severe disruptions to timetabled services for several weeks.

The cause of the damage to the seawall cannot be determined exactly, but is more than likely to have come about through a combination of wave impact forces on the front face (loosening and dislodging blocks), excessive wave overtopping (leading to ponding of water and significant seaward forces on the blocks due to the resulting head during the wave trough phase of the waves), and scour of the beach at the base of the wall (reducing wave breaking and allowing larger waves to reach the wall). The Venn diagram for this situation is shown in Figure 7.7.

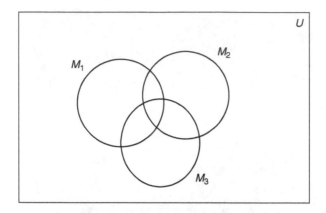

Figure 7.7 Venn diagram for the case when none of the three failure modes are mutually exclusive.

As before, let $P(M_1|M_3) = 0.4$. Furthermore, let $P(M_2|M_3) = 0.5$, $P(M_2|M_1M_3) = 0.8$ and $P(M_1M_2) = 0.0002$. Thus:

$$P(M_2M_3) = P(M_2|M_3)P(M_3) = 0.5 \times 0.01 = 0.005;$$
$$P(M_1M_3) = P(M_1|M_3)P(M_3) = 0.4 \times 0.01 = 0.004; \text{and}$$
$$P(M_2M_1M_3) = P(M_2|M_1M_3)P(M_1|M_3)P(M_3) = 0.8 \times 0.04 \times 0.01$$
$$= 0.00032$$

In this case the probability of failure due to any combination of failure modes is

$$P_f = P\{M_1 \cup M_2 \cup M_3\} = \sum_{i=1}^{3} P(M_i) - P(M_1M_2) - P(M_2M_3)$$
$$- P(M_1M_3) + P(M_1M_2M_3) = 0.02908$$

Consequently, the probability that no failure occurs in a year is $1 - 0.02908 = 0.97092$. Additionally, the probabilities of other events may be computed. For example:

a   The probability of the simultaneous occurrence of all three modes = $P(M_1M_2M_3) = 0.00032$.
b   The probability of the simultaneous occurrence of overtopping and armour damage = $P(M_1M_2) = 0.0002$.
c   The probability of the simultaneous occurrence of overtopping and scour = $P(M_1M_3) = 0.004$.

d The probability of the simultaneous occurrence of armour damage and scour $= P(M_2M_3) = 0.005$.

e The probability of failure due to overtopping and armour damage given that scour has occurred $= P(M_1M_2|M_3) = P(M_1M_2M_3)/P(M_3) = 0.00032/0.01 = 0.032$.

While the theory of probability allows one to calculate such probabilities, the more difficult practical aspect is estimating the joint and conditional probability of multiple failure modes with any degree of certainty. Figure 7.8 shows the same section of seawall after repairs were completed.

7.2.4 Multiple structures with multiple-failure mechanisms

Many coastal defence systems consist of multiple structures which can fail in various ways. A straightforward reliability analysis is further complicated if, say, the risk of flooding or erosion is required, as this introduces a spatial element to the considerations. For example, a set of defences may protect an area from flooding. Various assets within the protected area may be at different levels or have lesser or greater resilience to the depth of flooding than others. These factors can dramatically modify the picture of risk, as opposed to the risk of a structure not performing. Such calculations require flood routing calculations coupled with information on land use, characteristics of individual properties and so on, which is beyond the scope of this book.

Figure 7.8 Completed repairs of the Teignmouth–Dawlish seawall (courtesy of Network Rail).

Figure 7.9 Area affected by failure of one element of flood defence.

However, the process can be illustrated in a simple manner (Figures 7.9 and 7.10). Consider the area at risk from the failure of a particular flood defence, shown in Figure 7.9; flood routing calculations will determine the extent and depth of flooding, and in the more sophisticated versions duration as well. However, this is not the whole story if we are considering the

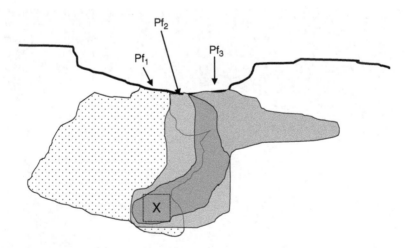

Figure 7.10 Flood risk at a location 'X' arising from failure of individual flood defence elements. The failure of more than one element of the flood defence system may lead to flooding at a fixed point behind the defence line, so the total probability of flooding at a point must take this into account.

risk to a particular property. Figure 7.10 shows the areas that will be inundated corresponding to the failure of three separate sections of defences. The risk of flooding at a property is the combination of risks arising from the failure of each element in the flood defence. In this type of situation it is common to assume that the elements are independent so that bounds on the total risk can be obtained from Table 7.1. However, the probabilities of failure of each element would first of all have to have been defined through an analysis such as that outlined in Section 7.2.3.

7.3 Flood defence schemes

In this section an example of a practical application of a probabilistic approach to flood risk assessment is described. The main issues encountered in this example can be summarised as follows:

- Large numbers of structures to be assessed (over 2000);
- Multiple failure modes;
- Variable quality of data;
- Quantifying the impact of a 'major storm event'.

Example 7.4. *National scale assessment of flood risk.* In the mid-1990s the UK insurance industry commissioned a study to assess the risk of coastal flooding for England and Wales. The study was commissioned by the Association of British Insurers (ABI), which is an umbrella organisation for its membership – insurance companies. The ABI had been aware of the rising cost of flood claims and wished to satisfy itself that it had sufficient reserves to cover the likely scale of claims should a major storm occur. The study covered sea and tidal defences on the open coast and within the Thames, Humber, Tees and Severn estuaries. In total there were over 1300 km of sea defences, with over 2000 different defence elements.

The main sources of information for the assessment were:

- The UK National Sea Defence Survey (NSDS), undertaken in the early 1990s, which was essentially a condition survey to guide maintenance and capital scheme requirements over a 5–10 year period;
- Areas defended against flooding as outlined in the NSDS;
- Water-level data, in the form of a 30-year time series for 16 locations and annual maxima for the remaining locations, from the Proudman Oceanographic Laboratory;
- Deep-water wave hindcasts, in the form of a 30-year time series at 3-hourly intervals, from the North European Storms Study at 32 locations;

- Additional wave time series for the Thames and Severn estuaries from the UK Meteorological Office wave model (Golding 1983), and for the Humber estuary from local wind records and wave hindcasting;
- Bathymetric data, for wave transformation calculations, obtained from Admiralty Charts.

A detailed discussion of the methodology adopted can be found in Reeve and Burgess (1994), and discussions of the mapping procedures in Burgess and Reeve (1994) and Maddrell et al. (1995, 1996). Here, a brief description is given so that the basic concepts can be appreciated. The most important point is that the results were intended to identify areas of high, medium and low flood risk. High risk was defined as an area vulnerable to an event with a 50-year return period, low risk as an area vulnerable to an event having a return period in excess of 200 years, with medium risk falling between these limits.

The methodology proceeded as follows:

1   Three modes of failure were considered: overflow, overtopping and toe failure.
2   Water-level data were analysed to determine extreme values corresponding to 50-, 100- and 200-year return periods (see Chapter 4).
3   Wave time series were transformed to inshore points using a spectral ray model (see e.g., Reeve et al. 2004), which accounted for refraction, shoaling and breaking.
4   Time series of inshore waves were analysed to determine extreme significant wave heights and corresponding wave periods (see Chapter 4).
5   Information on beach levels was used to define extreme (minimum) beach levels at each defence element.
6   The extreme beach levels were used in two ways: to determine the breaking wave limit at the structure and to determine the stability of the toe of the structure.
7   Calculations were performed using the 50-, 100- and 200-year extreme beach, wave and water levels to determine whether the structure could withstand these.
8   To determine the overall risk, overflow and overtopping were considered to be independent of toe failure but to have some dependence through water level. To provide a robust estimation of the risk, the combined probability of overtopping and overflow risk was taken as the larger of the two (i.e., complete dependence), and this was then combined with the probability of toe failure assuming independence.

Figure 7.11 shows the distribution of offshore wave and water-level locations that were used to define the near-shore wave and water-level conditions at almost 160 near-shore points. Each near-shore point was used to determine the wave conditions at approximately 10–20 defence lengths.

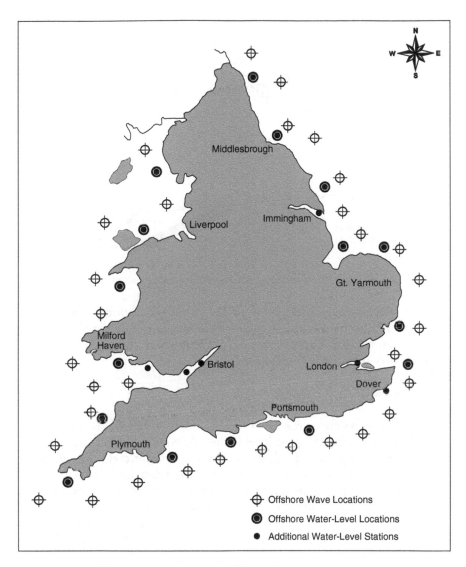

Figure 7.11 Offshore wave and water-level locations.

One of the main difficulties faced in this study was the very variable degree of information on the beach levels and the structures. Figure 7.12 illustrates some examples of defence profiles abstracted from the NSDS, and shows that in some cases further effort was required to define crest levels and materials. Indeed, almost 50% of all defences listed in the NSDS had poor quality (or no) data regarding the structure toe or beach levels.

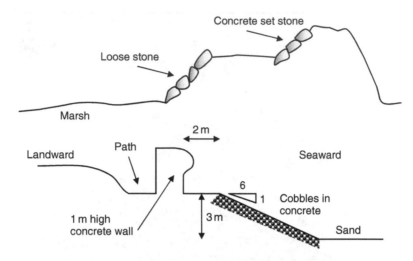

Figure 7.12 Examples of information on flood defences extracted from the NSDS.

The NSDS also gave information on the current condition of the defence. This was used to modify the strength of the structure in a simple way, not dissimilar to a Level 1 approach. So, for example, in the case of overflow, the reliability function is defined as:

$$G_{\text{overflow}} = l_e - l_{mw} \tag{7.3}$$

where l_e is the effective crest level of the structure and l_{mw} is the extreme water level. The effective crest level is the NSDS quoted level reduced by a factor based on engineering judgement to account for its condition (typically between 0.0 m for a new structure and 0.5 m for a damaged or ageing structure). The overtopping rate during the extreme storm (50-, 100- or 200-year event) was determined using Owen's formula (Equation 6.22), but adjusting the crest level as in Equation (7.3) to account for the condition of the defence. This was compared to damage threshold overtopping rates depending on the construction of the defence, as recommended in Owen (1980). If the computed overtopping rate exceeded the damage threshold, failure was considered to have occurred. Toe failure was defined as the probability that the beach level at the toe of the structure fell below a critical level required for the stability of the structure:

$$G_{\text{toe failure}} = l_b - l_c \tag{7.4}$$

where l_b is the annual minimum beach level and l_c is the critical beach level. Beach levels were described by a Gumbel distribution (for minima). In cases where no or insufficient beach measurements existed, a qualitative method

based on engineering judgement and historical records was used to cate-
gorise the risk of failure due to toe erosion as being high, medium or low.
The areas at risk of flooding were determined from the results of the defence
failure. Potential flood areas were established using a geographical information system and locating (from Ordnance Survey maps) the likely effect of
controlling structures such as embankments, roads and railways. The potential flood areas were then mapped onto postcode areas, which are used by
insurers as the first step in defining premium levels. Further results of this
project and a follow-on project have been reported in Maddrell et al. (1997,
1998). Even with the conservative assumptions mentioned above, the result
of the analysis was that less than 3% of the area protected by flood defences
was at high risk.

It is interesting to note that, in a more recent assessment, Oakes (2002)
reported that approximately 8% of England and Wales is at non-negligible
risk of flooding. Of this area, some 15% (or 2500 km²) is at risk of direct
flooding by the sea, potentially affecting 1.9 million homes.

Three important points to note are:

- To find a robust estimate of flood risk in such a wide-ranging project
  it is necessary to have a large data set, and a strategy for dealing with
  variable data quality.
- At one or more stages, it is quite likely that there will be insufficient
  information to determine statistical independence or mutual exclusivity,
  or otherwise. A pragmatic decision has therefore to be made, and the
  natural choice is to assume a 'worst-case' situation that gives a higher
  probability of failure.
- This risk assessment is a 'snapshot' at a particular point in time, and the
  risk of flooding will be a function of time, changing as environmental
  conditions evolve and structures weather and deteriorate.

7.4 Time-varying risk

While the various methods of estimating the risk of flooding or erosion have
been at the forefront of much of this book, another underlying theme should
also have been evident: that is, the risk of failure of a hydraulic structure or
coastal defence is a function of time because of the nature of the hydrodynamic loading and the behaviour of the structural materials. The techniques
mentioned in Chapters 5 and 6 provide examples of how time variation can
be incorporated to a degree. The most natural way to model time variation
is by using stochastic variables, particularly if the statistics of the variables in
question are nonstationary. However, to properly specify the characteristics
of a stochastic variable requires extensive observations. These are often not
available or are prohibitively expensive to collect; so alternative approaches
such as those described earlier are necessary if estimates are to be provided
to answer planning and design questions.

Further reading

Abernethy, C. L. and Gilbert, G., 1975. *Refraction of Wave Spectra*, Hydraulics Research Station, Wallingford Report INT117, May 1975.

Buijs, F. A., Sayers, P. B., Hall, J. W. and van Gelder, P. H. A. J. M., 2009. Time-dependent reliability analysis of anchored sheet pile walls, in *Flood Risk Management: Research and Practice* (Ed. P. Samuels), Taylor & Francis Group, London, pp. 503–513.

Peters, D. J. and Shaw, C. J., 1993. Modelling the North Sea through the North European Storm Study, Proceedings of 25th Offshore Technology Conference, May 1993, Houston, Texas, USA, Vol. 1, pp. 479–494.

Van Gelder, P. H. A. J. M., Buijs, F. A., ter Horst, W., Kanning, W., Van, C. M., Rajabalinejad, M., de Boer, E., Gupta, S., Shams, R., van Erp, N., Gouldby, B., Kingston, G., Sayers, P., Wills, M., Kortenhaus, A. and Lambrecht, H.-J., 2009. Reliability analysis of flood defence structures and systems in Europe, in *Flood Risk Management: Research and Practice* (Ed. P. Samuels), Taylor & Francis Group, London, pp. 603–611.

Epilogue

When visiting a site during calm, sunny conditions it can be tempting to think that severe storms are so rare that there is little need to consider their occurrence during construction or over the life of a scheme. A recent paper[1] illustrates the dangers of such a train of thought. The extension of the port of Gijón in the north of Spain was planned and designed meticulously, using the latest probabilistic approaches.[2] The construction took place over 3 years, predominantly during the summer months. During winter, the partly completed structure was capped with protective layers designed to have a useful life of 2 years and a failure probability of 0.5. Despite this, a very extreme storm (estimated at greater than 1 in 1000 year return period) occurred during the construction phase and caused damage to the structure. The extremity of the storm was due to the combination of wave height, period, direction and duration. Significant wave heights of up to 6.63 m, with a maximum wave height of 11.20 m, were observed, together with peak periods of 18 seconds, incidence directions from north-northeast, maximum sea levels of 5.50 m and a duration of 3 days. Fortunately, the extent of the damage was limited and was repaired without any great delay to the project. Had there been no protective layers the extent of the damage would have been much, much greater. Indeed, the careful calculation of probabilities and adherence to agreed guidelines meant that a *force majeure* was declared and the costs of damage were covered by insurance. This recent and real-life case illustrates that where water, and particularly the sea, is concerned, one cannot take anything for granted.

Returning to the issue of whether to use probabilistic methods or not – my advice to those designing flood defences would be, never, ever 'feel lucky'. Collect the measurements; perform the analyses and reliability calculations

1 Diaz Rato, J. L., Moyano Retamero, J. and de Miguel Riestra, M., 2009. Extension to the port of Gijón, *Proc ICE Maritime Engineering*, 161(MA4): 143–152.
2 Puertos del Estado, 2002. *Recommendations for Maritime Structures, ROM 0.0, General procedure and requirements in the design of harbour and maritime structures, Part 1.* Ministerio de Fomento, Madrid, Spain. ISBN 84-88975-31-7.

so you can make a sound engineering judgement of the risks involved. One can then weigh the probabilities of occurrence against the consequences and make an informed decision as to the appropriate level of protection required during both construction and for the completed scheme, allowing for the uncertainties inherent in the whole procedure.

Appendix A: normal distribution tables

Table A1 Standard normal distribution function $\Phi(z)$

Z	$\Phi(z)$	Z	$\Phi(z)$	Z	$\Phi(z)$
0.00	0.50000	1.00	0.84134	2.00	0.97724
0.02	0.50797	1.02	0.84613	2.02	0.97830
0.04	0.51595	1.04	0.85083	2.04	0.97932
0.06	0.52392	1.06	0.85542	2.06	0.98030
0.08	0.53188	1.08	0.85992	2.08	0.98123
0.10	0.53982	1.10	0.86433	2.10	0.98213
0.12	0.54775	1.12	0.86864	2.12	0.98299
0.14	0.55567	1.14	0.87285	2.14	0.98382
0.16	0.56355	1.16	0.87697	2.16	0.98461
0.18	0.57142	1.18	0.88099	2.18	0.98537
0.20	0.57925	1.20	0.88493	2.20	0.98609
0.22	0.58706	1.22	0.88876	2.22	0.98679
0.24	0.59483	1.24	0.89251	2.24	0.98745
0.26	0.60256	1.26	0.89616	2.26	0.98808
0.28	0.61026	1.28	0.89972	2.28	0.98869
0.30	0.61791	1.30	0.90319	2.30	0.98927
0.32	0.62551	1.32	0.90658	2.32	0.98982
0.34	0.63307	1.34	0.90987	2.34	0.99035
0.36	0.64057	1.36	0.91308	2.36	0.99086
0.38	0.64802	1.38	0.91620	2.38	0.99134
0.40	0.65542	1.40	0.91924	2.40	0.99180
0.42	0.66275	1.42	0.92219	2.42	0.99223
0.44	0.67003	1.44	0.92506	2.44	0.99265
0.46	0.677724	1.46	0.92785	2.46	0.99305
0.48	0.68438	1.48	0.93056	2.48	0.99343
0.50	0.69146	1.50	0.93319	2.50	0.99379
0.52	0.69846	1.52	0.93574	2.52	0.99413
0.54	0.7054	1.54	0.93821	2.54	0.99445
0.56	0.71226	1.56	0.94062	2.56	0.99476
0.58	0.71904	1.58	0.94294	2.58	0.99505

Table A1 (Continued)

Z	$\Phi(z)$	Z	$\Phi(z)$	Z	$\Phi(z)$
0.60	0.72574	1.60	0.94520	2.60	0.99533
0.62	0.73237	1.62	0.94738	2.62	0.99560
0.64	0.73891	1.64	0.94949	2.64	0.99585
0.66	0.74537	1.66	0.95154	2.66	0.99609
0.68	0.75174	1.68	0.95352	2.68	0.99631
0.70	0.75803	1.70	0.95543	2.70	0.99653
0.72	0.76423	1.72	0.95728	2.72	0.99673
0.74	0.77035	1.74	0.95907	2.74	0.99692
0.76	0.77637	1.76	0.96079	2.76	0.99710
0.78	0.78230	1.78	0.96246	2.78	0.99728
0.80	0.78814	1.80	0.96406	2.80	0.99744
0.82	0.79389	1.82	0.96562	2.82	0.99759
0.84	0.79954	1.84	0.96711	2.84	0.99774
0.86	0.80510	1.86	0.96855	2.86	0.99788
0.88	0.81057	1.88	0.96994	2.88	0.99801
0.90	0.81593	1.90	0.97128	2.90	0.99813
0.92	0.82121	1.92	0.97257	2.92	0.99824
0.94	0.82639	1.94	0.97381	2.94	0.99834
0.96	0.83147	1.96	0.97500	2.96	0.99846
0.98	0.83645	1.98	0.97614	2.98	0.99855
1.00	0.84134	2.00	0.97724	3.00	0.99865

Table A2 Standard normal distribution function
$\Phi(z)$ for larger values of z

Z	$\Phi(z)$
3.00	0.99865010
3.05	0.99885579
3.10	0.99903239
3.15	0.99918364
3.20	0.99931286
3.25	0.99942297
3.30	0.99951657
3.35	0.99959594
3.40	0.99966307
3.45	0.99971970
3.50	0.99976737
3.55	0.99980738
3.60	0.99984089
3.65	0.99986887
3.70	0.99989220
3.75	0.99991158
3.80	0.99992765

Z	$\Phi(z)$
3.85	0.99994094
3.90	0.99995190
3.95	0.99996092
4.00	0.99996832
4.05	0.99997439
4.10	0.99997934
4.15	0.99998337
4.20	0.99998665
4.25	0.99998931
4.30	0.99999146
4.35	0.99999319
4.40	0.99999458
4.45	0.99999570
4.50	0.99999660
4.55	0.99999731
4.60	0.99999788
4.65	0.99999834
4.70	0.99999869
4.75	0.99999898
4.80	0.99999920
4.85	0.99999938
4.90	0.99999952
4.95	0.99999962
5.00	0.9999997133

Table A3 Standard normal density function values, $\varphi(z)$

z	0.00	0.01	0.02	0.03	0.04	0.05	0.06	0.07	0.08	0.09
0.0	0.3989	0.3989	0.3989	0.3988	0.3986	0.3984	0.3982	0.3980	0.3977	0.3973
0.1	0.3970	0.3965	0.3961	0.3956	0.3951	0.3945	0.3939	0.3932	0.3925	0.3918
0.2	0.3910	0.3902	0.3984	0.3885	0.3876	0.3867	0.3857	0.3847	0.3836	0.3825
0.3	0.3814	0.3802	0.3790	0.3778	0.3765	0.3752	0.3739	0.3725	0.3712	0.3697
0.4	0.3683	0.3668	0.3653	0.3637	0.3621	0.3605	0.3589	0.3572	0.3555	0.3538
0.5	0.3521	0.3503	0.3485	0.3467	0.3448	0.3429	0.3410	0.3361	0.3372	0.3352
0.6	0.3332	0.3312	0.3292	0.3271	0.3251	0.3230	0.3209	0.3187	0.3166	0.3144
0.7	0.3123	0.3101	0.3079	0.3056	0.3034	0.3011	0.2989	0.2966	0.2943	0.2920
0.8	0.2897	0.2874	0.2850	0.2827	0.2803	0.2780	0.2756	0.2732	0.2709	0.2685
0.9	0.2661	0.2637	0.2613	0.2589	0.2565	0.2541	0.2516	0.2492	0.2468	0.2444
1.0	0.2420	0.2396	0.2371	0.2347	0.2323	0.2299	0.2275	0.2251	0.2227	0.2203
1.1	0.2179	0.2155	0.2131	0.2107	0.2083	0.2059	0.2036	0.2012	0.1989	0.1965
1.2	0.1942	0.1919	0.1895	0.1872	0.1849	0.1826	0.1804	0.1781	0.1758	0.1736
1.3	0.1714	0.1691	0.1669	0.1647	0.1626	0.1604	0.1582	0.1561	0.1539	0.1518
1.4	0.1497	0.1476	0.1456	0.1435	0.1415	0.1394	0.1374	0.1354	0.1334	0.1315
1.5	0.1295	0.1276	0.1257	0.1238	0.1219	0.1200	0.1182	0.1163	0.1145	0.1127
1.6	0.1109	0.1092	0.1074	0.1057	0.1040	0.1023	0.1006	0.0989	0.0973	0.0957
1.7	0.0940	0.0925	0.0909	0.0893	0.0878	0.0863	0.0848	0.0833	0.0818	0.0804
1.8	0.0790	0.0775	0.0761	0.0748	0.0734	0.0721	0.0707	0.0694	0.0681	0.0669
1.9	0.0656	0.0644	0.0632	0.0620	0.0608	0.0596	0.0584	0.0573	0.0562	0.0551

z	0.00	0.01	0.02	0.03	0.04	0.05	0.06	0.07	0.08	0.09
2.0	0.0540	0.0529	0.0519	0.0508	0.0498	0.0488	0.0478	0.0468	0.0459	0.0449
2.1	0.0440	0.0431	0.0422	0.0413	0.0404	0.0396	0.0387	0.0379	0.0371	0.0363
2.2	0.0355	0.0347	0.0339	0.0332	0.0325	0.0317	0.0310	0.0303	0.0297	0.0290
2.3	0.0283	0.0277	0.0270	0.0264	0.0258	0.0252	0.0246	0.0241	0.0235	0.0229
2.4	0.0224	0.0219	0.0213	0.0208	0.0203	0.0198	0.0194	0.0189	0.0184	0.0180
2.5	0.0175	0.0171	0.0167	0.0163	0.0158	0.0154	0.0151	0.0147	0.0143	0.0139
2.6	0.0136	0.0132	0.0129	0.0126	0.0122	0.0119	0.0116	0.0113	0.0110	0.0107
2.7	0.0104	0.0101	0.0099	0.0096	0.0093	0.0091	0.0088	0.0086	0.0084	0.0081
2.8	0.0079	0.0077	0.0075	0.0073	0.0071	0.0069	0.0067	0.0065	0.0063	0.0061
2.9	0.0060	0.0058	0.0056	0.0055	0.0053	0.0051	0.0050	0.0048	0.0047	0.0046
3.0	0.0044	0.0043	0.0042	0.0040	0.0039	0.0038	0.0037	0.0036	0.0035	0.0034
3.1	0.0033	0.0032	0.0031	0.0030	0.0029	0.0028	0.0027	0.0026	0.0025	0.0025
3.2	0.0024	0.0023	0.0022	0.0022	0.0021	0.0020	0.0020	0.0019	0.0018	0.0018
3.3	0.0017	0.0017	0.0016	0.0016	0.0015	0.0015	0.0014	0.0014	0.0013	0.0013
3.4	0.0012	0.0012	0.0012	0.0011	0.0011	0.0010	0.0010	0.0010	0.0009	0.0009
3.5	0.0009	0.0008	0.0008	0.0008	0.0008	0.0007	0.0007	0.0007	0.0007	0.0006
3.6	0.0006	0.0006	0.0006	0.0005	0.0005	0.0005	0.0005	0.0005	0.0005	0.0004
3.7	0.0004	0.0004	0.0004	0.0004	0.0004	0.0004	0.0003	0.0003	0.0003	0.0003
3.8	0.0003	0.0003	0.0003	0.0003	0.0003	0.0002	0.0002	0.0002	0.0002	0.0002
3.9	0.0002	0.0002	0.0002	0.0002	0.0002	0.0002	0.0002	0.0002	0.0001	0.0001
4.0	0.0001	0.0001	0.0001	0.0001	0.0001	0.0001	0.0001	0.0001	0.0001	0.0001

Table A4 Complement of the standard normal distribution function, $1 - \Phi(z)$

z	0.00	0.01	0.02	0.03	0.04	0.05	0.06	0.07	0.08	0.09
0.0	0.5000	0.4960	0.4920	0.4880	0.4840	0.4801	0.4761	0.4721	0.4681	0.4641
0.1	0.4602	0.4562	0.4522	0.4483	0.4443	0.4404	0.4364	0.4325	0.4286	0.4247
0.2	0.4207	0.4168	0.4129	0.4090	0.4052	0.4013	0.3974	0.3936	0.3897	0.3859
0.3	0.3821	0.3783	0.3745	0.3707	0.3669	0.3632	0.3594	0.3557	0.3520	0.3483
0.4	0.3446	0.3409	0.3372	0.3336	0.3300	0.3264	0.3228	0.3192	0.3156	0.3121
0.5	0.3085	0.3050	0.3015	0.2981	0.2946	0.2912	0.2877	0.2843	0.2810	0.2776
0.6	0.2743	0.2709	0.2676	0.2643	0.2611	0.2578	0.2546	0.2514	0.2483	0.2451
0.7	0.2420	0.2389	0.2358	0.2327	0.2296	0.2266	0.2236	0.2206	0.2177	0.2148
0.8	0.2119	0.2090	0.2061	0.2033	0.2005	0.1977	0.1949	0.1922	0.1894	0.1867
0.9	0.1841	0.1814	0.1788	0.1762	0.1736	0.1711	0.1685	0.1660	0.1635	0.1611
1.0	0.1587	0.1562	0.1539	0.1515	0.1492	0.1469	0.1446	0.1423	0.1401	0.1379
1.1	0.1357	0.1335	0.1314	0.1292	0.1271	0.1251	0.1230	0.1210	0.1190	0.1170
1.2	0.1151	0.1131	0.1112	0.1093	0.1075	0.1056	0.1038	0.1020	0.1003	0.0985
1.3	0.0968	0.0951	0.0934	0.0918	0.0901	0.0885	0.0869	0.0853	0.0838	0.0823
1.4	0.0808	0.0793	0.0778	0.0764	0.0749	0.0735	0.0721	0.0708	0.0694	0.0681
1.5	0.0668	0.0655	0.0643	0.0630	0.0618	0.0606	0.0594	0.0582	0.0571	0.0559
1.6	0.0548	0.0537	0.0526	0.0516	0.0505	0.0495	0.0485	0.0475	0.0465	0.0455
1.7	0.0446	0.0436	0.0427	0.0418	0.0409	0.0401	0.0392	0.0384	0.0375	0.0367
1.8	0.0359	0.0351	0.0344	0.0336	0.0329	0.0322	0.0314	0.0307	0.0301	0.0294
1.9	0.0287	0.0281	0.0274	0.0268	0.0262	0.0256	0.0250	0.0244	0.0239	0.0233

z	0.00	0.01	0.02	0.03	0.04	0.05	0.06	0.07	0.08	0.09
2.0	0.0228	0.0222	0.0217	0.0212	0.0207	0.0202	0.0197	0.0192	0.0188	0.0183
2.1	0.0179	0.0174	0.0170	0.0166	0.0162	0.0158	0.0154	0.0150	0.0146	0.0143
2.2	0.0139	0.0136	0.0132	0.0129	0.0125	0.0122	0.0119	0.0116	0.0113	0.0110
2.3	0.0107	0.0104	0.0102	0.0099	0.0096	0.0094	0.0091	0.0089	0.0087	0.0084
2.4	0.0082	0.0080	0.0078	0.0075	0.0073	0.0071	0.0069	0.0068	0.0066	0.0064
2.5	0.0062	0.0060	0.0059	0.0057	0.0055	0.0054	0.0052	0.0051	0.0049	0.0048
2.6	0.0047	0.0045	0.0044	0.0043	0.0041	0.0040	0.0039	0.0038	0.0037	0.0036
2.7	0.0035	0.0034	0.0033	0.0032	0.0031	0.0030	0.0029	0.0028	0.0027	0.0026
2.8	0.0026	0.0025	0.0024	0.0023	0.0023	0.0022	0.0021	0.0021	0.0020	0.0019
2.9	0.0019	0.0018	0.0018	0.0017	0.0016	0.0016	0.0015	0.0015	0.0014	0.0014
3.0	0.0013	0.0013	0.0013	0.0012	0.0012	0.0011	0.0011	0.0011	0.0010	0.0010
3.1	0.0010	0.0009	0.0009	0.0009	0.0008	0.0008	0.0008	0.0008	0.0007	0.0007
3.2	0.0007	0.0007	0.0006	0.0006	0.0006	0.0006	0.0006	0.0005	0.0005	0.0005
3.3	0.0005	0.0005	0.0005	0.0004	0.0004	0.0004	0.0004	0.0004	0.0004	0.0003
3.4	0.0003	0.0003	0.0003	0.0003	0.0003	0.0003	0.0003	0.0003	0.0003	0.0002
3.5	0.0002	0.0002	0.0002	0.0002	0.0002	0.0002	0.0002	0.0002	0.0002	0.0002
3.6	0.0002	0.0002	0.0001	0.0001	0.0001	0.0001	0.0001	0.0001	0.0001	0.0001
3.7	0.0001	0.0002	0.0001	0.0001	0.0001	0.0001	0.0001	0.0001	0.0001	0.0001
3.8	0.0001	0.0001	0.0001	0.0001	0.0001	0.0001	0.0001	0.0001	0.0001	0.0001
3.9	0.0000	0.0000	0.0000	0.0000	0.0000	0.0000	0.0000	0.0000	0.0000	0.0000

Appendix B: flow diagrams for reliability analysis

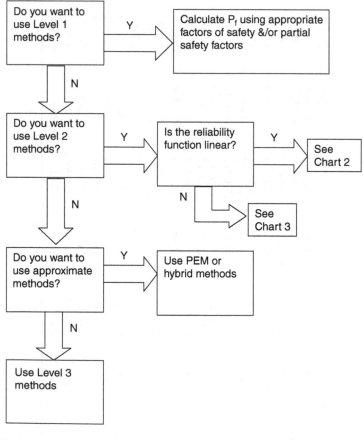

Chart 1

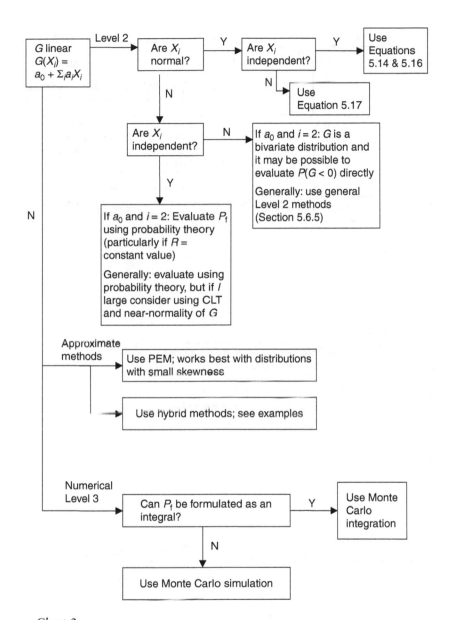

Chart 2

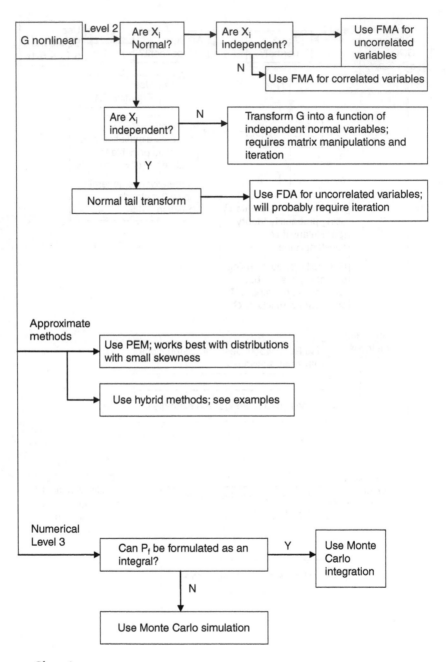

Chart 3

References

Abdo, R. and Rackwitz, R., 1990. A new beta-point algorithm for large variable problems in time-invariant and time-variant problems. In A. Der Kiureghian and P. Thoft-Christensen (Eds) *Proceedings of the 3rd IFIP WG7.5 Working Conference on Reliability and Optimisation of Structural Systems*, Springer, Berlin, p. 112.

Abramowitz, M. and Stegun, I., 1964. *Handbook of Mathematical Functions*, Applied Mathematics Series, Vol. 55, National, Bureau of Standards, Washington (reprinted by Dover Publications, New York).

ASCE, 1982. *Failure of the Breakwater of Port Sines, Portugal*, Vicksburg, USA, ISBN0872622983, p. 287.

Benjamin, J. R. and Cornell, C. A., 1970. *Probability, Statistics, and Decisions for Civil Engineers*, McGraw-Hill Book Company, New York, NY.

Blockley, D. I., 1992. (Ed.) *Engineering Safety*, McGraw-Hill Book Company, London, UK.

Booij, N., Holthuijsen, L. H. and Ris, R. C., 1996. The SWAN wave model for shallow water. Proceedings of the 25th International Conference on Coastal Engineering, ASCE, Orlando, pp. 668–676.

Borgman, L. E. and Sheffner, N. W., 1991. Simulation of time sequences of wave height, period and direction. Technical report. US Army Corps of Engineers.

Brater, E. F. and King, H. W., 1976. *Handbook of Hydraulics*, 6th edition, McGraw-Hill Book Company, New York.

Breitung, K., 1984. Asymptotic approximations for multinormal integrals. *Journal of Engineering Mechanics, ASCE*, 110(3): 357–366.

Breitung, K., 1994. *Asymptotic Approximations for Probability Integrals*, Springer-Verlag, Berlin.

Burcharth, H. F., 1992. Reliability evaluation of a structure at sea. Proceedings of the Short Course on Design and Reliability of Coastal Structures, attached to the Proceedings of 23rd International Conference of Coastal Engineers, ASCE, Venice, pp. 21.1–21.48.

Burcharth, H. F., Liu, A., d'Angremond, K. and van der Meer, J. W., 2000. Empirical formula for breakage of Dolosse and Tetrapods. *Coastal Engineering*, 40(3): 183–206.

Burcharth, H. F. and Sorensen, J. D., 1998. Design of vertical wall caisson breakwaters using partial safety factors. Proceedings of 26th International Conference on Coastal Engineering, ASCE, Copenhagen, pp. 2138–2151.

Burgess, K. A. and Reeve, D. E., 1994. The development of a method for the assessment of sea defences and risk of flooding. Proceedings of 29th MAFF Conference of River and Coastal Engineers, Loughborough, pp. 5.3.1–5.3.12.

Cai, Y., Gouldby, B., Hawkes, P. and Dunning, P., 2008. Statistical simulation of flood variables: Incorporating short-term sequencing. *Journal of Flood Risk Management*, 1: 1–10.

CERC, 1984. *Shore Protection Manual.* Coastal Engineering Research Center, U.S. Corps of Engineers, Vicksburg.

Chadwick, A. and Morfett, J., 1999. *Hydraulics in Civil and Environmental Engineering*, 3rd edition, E & FN SPON, London, p. 600.

Chambers, J. M., Cleveland, W. S., Kleiner, B. and Tukey, P. A., 1983. *Graphical Methods for Data Analysis*, Duxbury, Boston, MA.

Chang, C. H., Tung, Y. K. and Yang, J. C., 1993. Evaluation of probability point estimate methods. *Applied Mathematical Modelling*, 19(2): 95–105.

Chow, V. T., 1959. *Open Channel Hydraulics*, McGraw-Hill Book Company, New York, NY.

CIAD, 1985. *Computer Aided Evaluation of the Reliability of a Breakwater Design*, Final Report, CIAD Project Group, July 1985.

CIRIA, 1977. *Rationalisation of Safety and Serviceability Factors in Structural Design*, Report 63, London.

CIRIA, 1996. *Beach Management Manual*, Report 153, London.

CIRIA and CUR, 1991. *Manual on the Use of Rock in Coastal and Shoreline Engineering*, CIRIA Special Publication 83/CUR Report 154, London.

Coles, S. G. and Tawn, J. A., 1991. Modelling extreme multivariate events. *Journal of the Royal Statistical Society: Series B*, 53: 377–392.

Coles, S. G. and Tawn, J. A., 1994. Statistical methods for multivariate extremes: an application to structural design. *Applied Statistics*, 43(1):1–48.

Council Directive (76/160/EEC), 1975. Council Directive of 8 December 1975 concerning the Quality of Bathing Water (76/160/EEC).

Council Directive (91/271/EEC), 1991. Council Directive of 21 May 1991 concerning urban waste water treatment (91/271/EEC).

Cunnane, C., 1978. Unbiased plotting positions – a review, *Journal of Hydrology*, 37: 205–222.

Dabees, M. and Kamphuis, W. J., 1998. ONELINE, a numerical model for shoreline change. Proceedings of 26th International Conference on Coastal Engineering, Copenhagen, Denmark, pp. 2668–2681.

Davison, A. C. and Smith, R. L., 1990. Models for exceedances over high thresholds. *Journal of the Royal Statistical Society: Series B*, 52: 393–442.

De Haan, L., 1976. Simple extremes: an elementary introduction, *Statistica Neerlandica*, 30: 161–172.

De Haan, L. and Resnick, S., 1977. Limit theory for multivariate sample extremes. *Z. Wahrscheinlichkeitstheorie v. Geb.*, 40: 317–329.

De Vriend, H. J., 2003. On the prediction of aggregated-scale coastal evolution. *Journal of Coastal Research*, 19(4): 757–759.

Dean, R. G. and Dalrymple, R. A., 2002. *Coastal Processes: With Engineering Applications*, Cambridge University Press, Cambridge.

DEFRA, 2006. *Risk Assessment for Coastal Erosion*, DEFRA Project FD2324, The Stationery Office, London.

Der Kiureghian, A. and Liu, P.-L., 1986. Structural reliability under incomplete probability information. *Journal of Engineering Mechanics*, 112(1): 85–104.

DETR, Environment Agency, Institute for Environment and Health, 2000. *Guidelines for Environmental Risk Assessment and Management*, The Stationery Office, London.

Ditlevsen, O., 1979. Narrow reliability bounds for structural systems. *Journal of Structural Mechanics*, 7: 435–451.

Ditlevsen, O. and Madsen, H. O., 1996. *Structural Reliability Methods*, John Wiley & Sons, Chichester.

Efron, B., 1979. Bootstrap methods: another look at the jack-knife. *Annals of Statistics*, 7: 1–26.

Efron, B. and Tibshirani, R. J., 1993. *An Introduction to the Bootstrap*, Chapman and Hall, New York, 436pp.

Environment Agency, 1996. *Risk Assessment for Sea and Tidal Defence Schemes*. Report 459/9/Y. Environment Agency, Bristol, UK.

Environment Agency, 2000. *Risk Assessment for Sea and Tidal Defence Schemes*, Final R & D Report, Report No. 459/9/Y. Environment Agency, Bristol, UK.

EPA, 2007. *Water Quality Standard Handbook*, EPA-823-B-94-005 (1st edition 1994; 2nd edition in June 2007).

Ewans, K. and Jonathan, P., 2006. Estimating extreme wave design criteria incorporating directionality. 9th International Workshop on Wave Hindcasting and Forecasting, Victoria, Canada.

Failure Analysis Associates, Inc., 1993. *Design News*, Issue 10th April 1993.

Fiessler, B., Neumann, H.-J. and Rackwitz, R., 1979. Quadratic limit states in structural reliability. *Journal of Engineering Mechanics, ASCE*, 105(4): 661–676.

Fisher, R. A. and Tippett, H. C., 1928. Limiting forms of the frequency distribution of the largest or smallest member of a simple. *Proceedings of the Cambridge Philosophical Society*, 24: 180–190.

Fréchet, M., 1927. Sur la loi de probabilité de l'écart maximum, *Annales de la Societé Polonaise de Mathématique (Cracow)*, 6: 93–117.

Galiatsatou, P. and Prinos, P., 2007. Bivariate models for extremes of significant wave height and period. An application to the Dutch Coast. Proceedings of the 2nd IMA Conference on Flood Risk Assessment, 2007, Plymouth, Institute of Mathematics & Its Applications, pp. 77–84.

Germann, U. and Zawadzki, I., 2002. Scale dependency of the predictability of precipitation from continental radar images. Part 1: description of the methodology. *Monthly Weather Review*, 130(12): 2859–2873.

Giles, R. V., 1962. *Schaum's Outline on Fluid Mechanics and Hydraulics*, McGraw-Hill Book Company, New York, NY.

Gnedenko, B. V., 1943. Sur la distribution limite du terme maximum d'une série aléatoire. *Annals of Mathematics*, 44: 423–453.

Golding, B. W., 1983. A wave prediction system for real-time sea state forecasting. *Journal Royal Meteorological Society*, 109: 393–416.

Gravens, M. B., Kraus, N. C. and Hanson, H., 1991. *GENESIS: Generalized Model for Simulating Shoreline Change*, Report 2, Workbook and System User's Manual. U. S. Army Corps of Engineers.

Grecu, M. and Krajewski, W. F., 2000. A large-sample investigation of statistical procedures for radarbased short term quantitative precipitation forecasting. *Journal of Hydrology*, 239(1–4): 69–84.

Greenwood, J. A., Landwehr, J. M., Matalas, N. C. and Wallis, J. R., 1979. Probability weighted moments: definition and relation to parameters of several distributions expressible in inverse form. *Water Resources Research*, 15(5): 1049–1054.

Grijm, W., 1961. Theoretical forms of shoreline. Proceedings of the 7th International Conference on Coastal Engineering, ASCE, Sydney, Australia, pp. 219–235.

Gumbel, E. J., 1954. *Statistical Theory of Extreme Values and Some Practical Applications*, Applied Mathematics Series Publication No. 33, National Bureau of Standards, Washington, DC.

Habib, E. H. and Mesellie, E. A., 2000. Stage-discharge relations for low-gradient tidal streams using data-driven models. *Journal of Hydraulic Engineering*, 132(5): 482–492.

Hall, J., Dawson, R., Sayers, P., Rosu, C., Chatterton, J. and Deakin, R., 2003. A methodology for national–scale flood risk assessment. *Proceedings of the Institution of Civil Engineers – Water Maritime Engineering*, 156(WM3): 235–247.

Hall, J., Meadowcroft, I. C., Lee, M. and Van Gelder, P. H. A. J. M., 2002. Stochastic simulation of episodic soft coastal cliff recession. *Coastal Engineering*, 46: 159–174.

Hames, D., 2006. Investigation of the joint probability of waves and high sea levels along the Cumbrian Coast, PhD Thesis, University of Plymouth.

Hames, D. and Reeve, D. E., 2007. The joint probability of waves and high sea levels in coastal defence. Proceedings of the 2nd IMA Conference on Flood Risk Assessment, 2007, Plymouth, Institute of Mathematics & Its Applications, pp. 97–106.

Hammersley, J. M. and Hanscomb, D. C., 1964. *Monte Carlo Methods*, Methuen, London.

Harr, M. E., 1987. *Reliability-Based Design in Civil Engineering*, McGraw-Hill, New York, NY.

Hasofer, A. M. and Lind, N. C., 1974. An exact and invariant first order reliability format. *Proc ASCE, Journal of the Engineering Mechanics Division*, 100, 111–121.

Hawkes, P. J., Gouldby, B. P., Tawn, J. A. and Owen, M. W., 2002. The joint probability of waves and water levels in coastal engineering design. *Journal of Hydraulic Research*, 40(3): 241–251.

Herzog, M. A. M., 1982. Behind the Sines Portugal breakwater failure. *Journal of Waterway, Port, Coastal and Ocean Engineering*, ASCE, 52(4): 64–67.

Hohenbichler, M., Gollwitzer, S., Kruse, W. and Rackwitz, R., 1987. New light on first- and second-order reliability methods. *Structural Safety*, 4: 267–284.

Hosking, J. R. M., 1990. L-moments: analysis and estimation of distribution using linear combinations of order statistics. *Journal of the Royal Statistical Society: Series B*, 52: 105–124.

http://www.fwr.org/wapug/wapugdes.pdf (accessed 15/12/08)

http://zebu.uoregon.edu/1999/ph161/l20.html (accessed 24/01/08)

HR Wallingford, 2000. The joint probability of waves and water levels: JOINSEA – A rigorous but practical new approach, HR Wallingford Report SR 537, Final Edition, May 2000.

HSE, 2002. *Comparative Evaluation of Minimum Structures and Jackets*, Offshore Technology Report 2001/062, HMSO, Norwich, p. 88.

Hudson, R. Y., 1959. Laboratory investigations of rubble-mound breakwaters. *Proc. ASCE*, 85(3): 93–121.

Jakeman, E., 1980. On the statistics of K-distributed noise. *Journal of Physics*, A13: 31–48.

James, A., 1993. *An Introduction to Water Quality Modelling*, 2nd edition, John Wiley & Sons, New York, NY, 234pp.

Johnson, P. A., 1992. Reliability-based pier scour engineering. *ASCE Journal of Hydraulic Engineering*, 118: 1344–1358.

Johnson, R. W. (Ed.), 1998. *Handbook of Fluid Dynamics*, CRC Press, Boca Raton, FL.

Kamphuis, J. W., 2001. *Introduction to Coastal Engineering and Management*, Advanced Series on Ocean Engineering, Vol. 16, World Scientific, Singapore.

Kjeldsen, T. R. and Jones, D. A., 2006. Prediction uncertainty in a median based index flood method using L-moments. *Water Resources Research*, 42: W07414, doi: 10.1029/2005WR004069.

Knuth, D. E., 1981. *Seminumerical Algorithms*, Vol. 2 of *The Art of Computer Programming*, 2nd edition, Addison-Wesley, Reading, MA.

Kottegoda, N. T. and Rosso, R., 1997. *Statistics, Probability, and Reliability for Civil and Environmental Engineers*, McGraw-Hill, New York, NY, 735pp.

Lang, S., 1974. *Calculus of Several Variables*, 3rd edition, Addison-Wesley, Reading, MA.

Larson, M., Capobianco, M., Jansen, H., Różyński, G., Southgate, H., Stive, M. and Wijnberg, K., 2003. Analysis and modelling of field data on coastal morphological evolution over yearly and decadal time scales: Part 1. Background and linear techniques. *Journal of Coastal Research*, 19(4): 760–775.

Larson, M., Hanson, H. and Kraus, N. C., 1987. *Analytical Solutions of the One-Line Model of Shoreline Change*, Technical Report CERC-87-15, USAE WES, Coastal Engineering Research Center, Vicksburg, Mississippi.

Larson, M., Hanson, H. and Kraus, N. C., 1997. Analytical solutions of one-line model for shoreline change near coastal structures. *Journal of Waterway, Port, Coastal, and Ocean Engineering*, ASCE, 123(4): 180–191.

Le Mèhautè, B. and Soldate, M., 1977. *Mathematical Modelling of Shoreline Evolution*. CERC Miscellaneous Report No. 77–10, USAE-WES, Coastal Engineering Research Center, Vicksburg, Mississippi.

Leadbetter, M. R., 1983. Extremes and local dependence in stationary sequences. *Z. Wahrscheinlichkeitsth.*, 65: 291–306.

Lee, E. M., Hall, J. W. and Meadowcroft, I. C., 2001. Coastal cliff recession: the use of probabilistic prediction methods. *Geomorphology*, 40: 253–269.

Li, B., 1994. An evolution equation for water waves. *Coastal Engineering*, 23: 227–242.

Li, K. S., 1992. Point estimate method for calculating statistical moments. *Journal of Engineering Mechanics*, 118(7): 1506–1511.

Li, Y., Simmonds, D. J. and Reeve, D. E., 2008. Quantifying uncertainty in extreme values of design parameters with resampling techniques. *Ocean Engineering*, 35(10): 1029–1038.

Linhart, H. and Zucchini, W., 1986. *Model Selection*, John Wiley & Sons, New York.

Lovejoy, S., Duncan, M. R. and Schertzer, D., 1996. Scalar multi-fractal radar observer's problem. *Journal of Geophysical Research (Atmospheres)*, 101(D21): 26479–26491.

Maddrell, R. J., Burgess, K. A., Deakin, R. and Mounsey, C., 1995. Identification of coastal flood areas in England and Wales for the insurance industry. Proceedings of 30th MAFF Conference of River and Coastal Engineers, University of Keele, July 1995, Keele, pp. 2.1.1–2.1.11.

Maddrell, R. J., Fleming, C. A. and Mounsey, C., 1997. Assessing coastal flood risks. Proceedings of 25th ICCE, Orlando, USA, Vol. IV, pp. 4339–4352.

Maddrell, R. J., Mounsey, C. and Burgess, K. A., 1996. Coastal flooding: Assessing the risk for the insurance industry, ICE Conference 1995, Brighton, UK.

Maddrell, R. J., Reeve, D. E. and Heaton, C. R., 1998. Establishing coastal flood risks from single storms and the distribution of the risk of structural failure, in *Coastlines, Structures and Breakwaters* (Ed. N. W. H. Allsop), Thomas Telford, London, pp. 20–33.

MAFF, 2000. *Flood and Coastal Defence Project Appraisal Guidance: Approaches to Risk*, Report FCDPAG4, UK Ministry of Agriculture, Fisheries and Food, London.

Matousek, M. and Schneider, J., 1976. *Untersuchungen zur Struktur des Sicherheitsproblems bei Bauwerken*, Bericht No. 59, Institut für Baustatik und Konstruktion, Eidgenössiche Technische Hochschule, Zurich.

Meadowcroft, I. C., von Lany, P. H., Allsop, N. W. H. and Reeve, D. E., 1995. Risk assessment for coastal and tidal defence schemes. Proceedings of the 24th International Conference on Coastal Engineering (ICCE '94), Kobe, Japan, 23–24 October 1994, pp. 3154–3166.

Meadowcroft, I. C., Reeve, D. E., Allsop, N. W. H., Deakin, R. and Cross, J. C., 1996a. Risk assessment for sea and tidal defences. Proceedings of 31st MAFF Conference of River and Coastal Engineers, Keele, July, pp. 3.3.1 – 3.3.12.

Meadowcroft, I. C., Reeve, D. E., Allsop, N. W. H., Diment, R. P. and Cross, J. C., 1996b. Development of new risk assessment procedures for coastal structures, in *Advances in Coastal Structures and Breakwaters* (Ed. J. E. Clifford), Thomas Telford, London, pp. 6–46.

Melchers, R. E., 1999. *Structural Reliability Analysis and Prediction*, 2nd edition, John Wiley & Sons, Chichester, p. 437.

Oakes, T., 2002. Flood and coastal defence and municipal engineers. *Municipal Engineer*, 151(4): 287–289.

O' Brien, G. L., 1987. Extreme values for stationary and Markov sequences. *Annals of Probability*, 15: 281–291.

Oumeraci, H., Kortenhaus, A., Allsop, N. W. H., de Groot, M. B., Crouch, R. S., Vrijling, J. K. and Voortman, H. G., 2001. *Probabilistic Design Tools for Vertical Breakwaters*, A A Balkema, Lisse, p. 373.

Owen, M. W., 1980. *Design of Seawalls Allowing for Wave Overtopping*, HR Wallingford Report EX924, UK.

Papoulis, A., 1984. *Probability, Random Variables and Stochastic Processes*, 2nd edition, Electrical Engineering Series, McGraw-Hill Book Company, Singapore, p. 576.

Pedrozo-Acuña, A., Reeve, D. E., Burgess, K. A., Hosking, A. and Loran, F., 2008. Risk assessment of coastal erosion in the United Kingdom, *Revista de Ingeniería Hidráulica en México*, IMTA, XXIII(4) (in Spanish).

Pedrozo-Acuña, A., Reeve, D. E. and Spivack, M. Ensemble prediction of shoreline position near a groyne, proceedings of ICCE 2008, Hamburg, ASCE (in press).

Pelnard-Considère, R., (1956). Essai de theorie de l'evolution des forms de rivages en plage de sable et de galets. Societe Hydrotechnique de France, Proceedings of the 4th Journees de l'Hydraulique, les Energies de la Mer, Question III, Rapport No. 1, pp. 289–298.

PIANC, 1992. *Analysis of Rubble Mound Breakwaters*, supplement to PIANC Bulletin No. 78/79, PIANC (Brussels).

Pickands, J., 1975. Statistical inference using extreme order statistics. *Annals of Statistics*, 3: 119–131.

Pickands, J., 1981. Multivariate extreme value distributions, *Bull.* I.S.I., XLIX (Book 2), pp. 859–878.

Press, W. H., Teukolsky, S. A., Vetterling, W. T. and Flannery, B. P., 2007. *Numerical Recipes*, 3rd edition, Cambridge University Press, Cambridge, p. 1235.

Quenouille, M., 1949. Approximation tests of correlation in time series. *Journal of the Royal Statistical Society: Series B*, 11: 18–84.

Radavich, J. F., 1980. 'Defects and Transformation Embrittlement', Reliability, Stress Analysis and Failure Prevention Methods in Mechanical Design, ASME.

Reeve, D. E., 1998. On coastal flood risk. *Journal of Waterway, Port, Coastal and Ocean Engineering, ASCE*, 124(5): 219–228.

Reeve, D. E., 2003. Probabilistic methods for flood risk assessment. In J. McKee Smith (Ed.) *Proceedings 28th ICCE*, July 2002, World Scientific, Cardiff, pp. 2324–2334.

Reeve, D. E., 2006. Explicit expression for beach response to non-stationary forcing near a groyne. *Journal of Waterway, Port, Coastal and Ocean Engineering, ASCE*, 132: 125–132.

Reeve, D. E. and Burgess, K. A., 1994. Assessment of coastal flood risk in England and Wales, IMA *Journal of Mathematics Applied in Business and Industry*, 5: 197–209.

Reeve, D. E., Chadwick, A. J. and Fleming, C. A., 2004. *Coastal Engineering: Processes, Theory and Design Practice*, SPON Press, Abingdon, p. 461.

Reeve, D. F., Horrillo-Caraballo, J. M. and Magar, V., 2008. Statistical analysis and forecasts of long-term sandbank evolution at Great Yarmouth, UK. *Estuarine, Coastal and Shelf Science*, 79(3): 387–399.

Reinen-Hamill, R. A., 1997. Numerical modelling of the Canterbury Regional Bight. The use of a shoreline evolution model for management and design purposes. Pacific Coasts and Ports '97: Proceedings of the 13th Australasian Coastal and Ocean Engineering Conference, Christchurch, New Zealand, pp. 359–364.

Resnick, S., 1987. *Extreme Values, Point Processes and Regular Variation*, Springer Verlag, Berlin.

Reynolds, O., 1894. *Philosophical Transactions of the Royal Society, London*, A186: 123.

Rosenblueth, E., 1975. Point estimates for probability moments, *National Academy of Sciences*, 72(10): 3812–3814.

Różyński, G., 2003. Data-driven modelling of multiple longshore bars and their interaction. *Coastal Engineering*, 48: 151–170.

Różyński, G. and Jansen, H., 2002. Modelling nearshore bed topography with principal oscillation patterns. *Journal of Waterway, Port, Coastal and Ocean Engineering, ASCE*, 128(5): 202–215.

Różyński, G., Larson, M. and Pruszak, Z., 2001. Forced and self-organised shoreline response for a beach in the southern Baltic Sea determined through singular spectrum analysis. *Coastal Engineering*, 43: 41–58, doi: 10.1016/S0378-3839(03)00024-3.

Scheffner, N. W. and Borgman, L. E., 1992. Stochastic time-series representation of wave data. *Journal of Waterway, Port, Coastal and Ocean Engineering, ASCE*, 118(4): 337–351.

Shankar, P. M., Dumane, V. A., Reid, J. M., Genis, V., Forsberg, F., Piccoli, C. W. and Goldberg, B. B., 2000. Use of the K-distribution for classification of breast masses. *Ultrasound in Medicine and Biology*, 26(9): 1503–1510.

Smith, R. L., 1986. Extreme value theory based on the r largest annual events, *Journal of Hydrology*, 86: 27–43.

Southgate, H. N., Wijnberg, K. M., Larson, M., Capobianco, M. and Jansen, H., 2003. Analysis of field data of coastal morphological evolution over yearly and decadal timescales. Part 2: Non-linear techniques. *Journal of Coastal Research*, 19(4): 776–789.

Tawn, J. A., 1988. Bivariate extreme value theory – models and estimation, *Biometrika*, 75: 397–415.

Thoft-Christensen, P. and Baker, M. J., 1982. *Structural Reliability Theory and Its Applications*, Springer, Berlin.

Thompson, P., 2009. Statistical techniques for extreme wave condition analysis in coastal design, PhD Thesis, University of Plymouth, p. 208.

Thompson, P., Reeve, D. E., Stander, J., Cai, Y. and Moyeed, R., 2008. Bayesian non-parametric quantile regression using splines for modelling wave heights, Proceedings of the European Conference on Flood Risk Management (FLOODrisk 2008), Oxford, UK, {*Flood Risk Management: Research and Practice – Samuels et al. (eds)*}, CRC Press, London, pp. 1055–1061.

Townend, I. H., 1994. Risk assessment of coastal defences. Proceedings of the Institution of Civil Engineers Water Maritime and Energy, 106: 381–384.

US Army Corps of Engineers, 1989. Water Levels and Wave Heights for Coastal Engineering Design, CECW-EH, Engineer Manual, 1110-2-1414.

Van der Meer, J. W., 1988. Deterministic and probabilistic design of breakwater armour layers. *Proc. ASCE, Journal of WPC and OE*, 114(1).

Vrijling, J. K., 1982. Probability design method, Eastern Scheldt Storm Surge Barrier. Proceedings of the Delta Barrier Symposium, Rotterdam, pp. 44–49.

Walker, A. C., 1981. Study and analysis of the first 120 failure cases. *Structural failures in Buildings*, Institution of Structural Engineers, London, pp. 15–39.

Williams, G. S. and Hazen, A., 1955. *Hydraulic Tables*, 3rd edition, John Wiley & Sons, New York, NY.

Winant, C. D., Inman, D. L. and Nordstrom, C. E., 1975. Description of seasonal beach changes using empirical eigen functions. *Journal of Geophysical Research*, 80(15): 1979–1986.

Wind, H. G., 1990. Influence functions. Proceedings of 21st International Conference on Coastal Engineering, Costal del Sol-Malaga, Spain, pp. 3281–3294.

Yen, B. C., 1992. Dimensionally homogeneous Manning's formula, *Journal of Hydraulic Engineering, ASCE*, 118(9): 1326–1332.

Yu, K., Lu, Z. and Stander, J., 2003. Quantile regression: application and current research areas. *Journal of the Royal Statistical Society: The Statistician*, 52(3): 331–350.

Yu, K. and Moyeed, R., 2001. Bayesian quantile regression. *Statistics and Probability Letters*, 54: 437–447.

Zacharioudaki, A. and Reeve, D. E., 2007. Explicit formulae of shoreline change under time-varying forcing. Proceedings of the Conference on River,

Coastal and Estuarine Morphodynamics: RCEM2007, Twente, Netherlands, pp. 1067–1074.

Zacharioudaki, A. and Reeve, D. E., 2008. Semi-analytical solutions of shoreline response to time varying wave conditions. *Journal of Waterway, Port, Coastal and Ocean Engineering, ASCE*, 134(5): 265–274.

Zawadzki, I., Morneau, J. and Laprise, R., 1994. Predictability of precipitation patterns: an operational approach. *Journal of Applied Meteorology*, 33: 1562–1571.

Zipparro, V. J. and Hasen, G., 1972. *Davis' Handbook of Applied Hydraulics* 4th edition, Tata McGraw-Hill, New York.

Index

Note: The notation "ff" along with locators refer to following folios.